Signals and Communication Technology

This series is devoted to fundamentals and applications of modern methods of signal processing and cutting-edge communication technologies. The main topics are information and signal theory, acoustical signal processing, image processing and multimedia systems, mobile and wireless communications, and computer and communication networks. Volumes in the series address researchers in academia and industrial R&D departments. The series is application-oriented. The level of presentation of each individual volume, however, depends on the subject and can range from practical to scientific.

Indexing: All books in "Signals and Communication Technology" are indexed by Scopus and zbMATH

For general information about this book series, comments or suggestions, please contact Mary James at mary.james@springer.com or Ramesh Nath Premnath at ramesh.premnath@springer.com.

Juraj Ďuďák • Gabriel Gašpar

Design and Implementation of Sensory Solutions for Industrial Environment

Utilizing 1-wire® Technology
in Industrial Solutions

 Springer

Juraj Ďuďák
Faculty of Materials Science and
Technology in Trnava
Slovak University of Technology in
Bratislava
Trnava, Slovakia

Gabriel Gašpar
Research Centre
University of Žilina
Žilina, Slovakia

ISSN 1860-4862   ISSN 1860-4870   (electronic)
Signals and Communication Technology
ISBN 978-3-031-30154-4   ISBN 978-3-031-30152-0   (eBook)
https://doi.org/10.1007/978-3-031-30152-0

This Springer imprint is published by the registered company Springer Nature Switzerland AG
The registered company address is: Gewerbestrasse 11, 6330 Cham, Switzerland

Foreword

For many years, in my group, we develop sensors, especially but not only for biomedical applications, specific analogue and digital electronics, power-minimized solutions, energy harvesting devices, wireless data transfer and new signal processing methods. Thus, I was very pleased as this book's authors told me some months ago that they would like to work temporarily in my team on sensors and signal processing methods for rehabilitation patients after skeletal operations and therapy. They brought many new technological approaches to my group, local solutions and new project ideas. Thus, after they asked me to write a foreword for their new book, which handles these and related themes, I agreed gladly.

The authors worked on five industrially oriented research and development projects that have been realized in a natural environment and confirmed the usability of new sensor technology, communication and signal processing. They decided to bring the methodology to students and practitioners through this book for use in study programs and industrial research and development. Starting didactically correctly with a detailed description of communication ways at all levels of hardware, software and protocols, the communications used and verified in projects are presented.

In the next chapter about the hardware present, the authors follow the chain, starting with technical basics and developing their basics of 1-wire technology. Then follows a detailed description of the design and realization of wireless communication combined with the wired network on the receiver side. Constructions for the natural environment proved in the real world are presented. Several measuring modules for a distributed sensor network are described in detail and evaluated: THB (Temperature-Humidity Board), TNZ (Tensometric Module) and standard Sensory Modules. The following chapter about software follows the same didactic structure. Application-specific self-developed software modules show how the authors solved industry-oriented projects for reasonable plausibility and reusability.

The authors have worked for a long time and work today in the field of sensor networks, wired and wireless data transfer in minimalized hardware and software structures to save energy, data transfer and signal processing and storing. The last chapter of this book introduces the most essential industrially implemented and evaluated use cases and projects. This book is the perfect collection of recipes for (not only) industrial applications well-grounded in communication, hardware and software, as well as signal processing basics up to realizations in the natural world. I will recommend it to my students and researchers in our projects.

In Ilmenau, 18. January 2023   *Univ.-Prof. Dr.-Ing.habil. Peter Husár*
                                                Head of Biosignal Processing Group
                                                Technische Universität Ilmenau
                                                Germany

Preface

I believe in the development of science, and therefore in the
development of all that is noble and beneficial to society.
Gen. PhDr. Milan Rastislav Štefánik

Sensor systems are an integral part of any control system, from the simplest ones in the form of individual sensors, through the more demanding solutions with their own additional hardware and software equipment, to advanced sensor systems, using both local and remote hardware and software resources, often highly demanding on computing power and technical equipment. Such systems are used throughout one country's economy while currently finding a place in the equipment of apartments and family houses as a part of the so-called "home automation." Sensor systems used in the role of industrial measuring systems must fulfill the following requirements:

- Data transmission must be protected against failures. The transmission is repeated at a specified time in the event of faulty transmission, while no data must be lost.
- If the subsystems work continuously and the control unit requests information only after specific time intervals, the transmission is minimized according to the needs of the control unit.
- The system must provide an interface for communication with other systems.

This publication presents a set of selected contributions in the field of sensory system design, hardware and software design, and data processing, as well as access to them.

The design of the sensory data collection system was based on requirements to minimize the resources needed for the system's functionality. Under the notion of resource, the following were considered: the communication protocol's complexity, the implementation's hardware complexity, and the physical layer used for mutual communication.

The objective of this publication was to define a complex sensor system for collecting industrial data. From the communication protocols' design (at different levels of communications), through the hardware design and communication infrastructure, up to access to data and their evaluation and the solution's integration into the industrial networks (e.g., Internet of Things or conventionally used industrial communication standards). In all the phases of this proposal, we relied on our professional experience, which we gained in the personal and academic sectors. In individual parts design, emphasis was always placed on the undemanding implementation while complying with all the requirements for the individual components. We placed a great emphasis on securing communication at the lowest level, i.e., between the sensor and the measuring unit and between the measuring unit and the control computer, because at this level, communication security is often absent, and data is transmitted without using a secure layer. We wanted to add the secure communication to the lowest communication layers because the sensor system's data can be used to control other processes. Here, a potential security threat appears that the state of the controlled process could be purposefully changed with the modified sensory data.

From the broader context point of view, the presented sensor system solution is designed in such a way that it can be implemented into the broader concept of Industry 4.0. The concept of Industry 4.0 is based on the fact that people, machines, devices, logistic systems, and products can directly communicate and cooperate. The reason is the use of a large amount of information that has not yet been processed to support faster and correct decision-making. Acquiring additional data, such as temperature, humidity, the intensity of light radiation, atmospheric pressure, and others, can substantially supplement complex information in the production process. With the help of this data, it is possible to obtain otherwise hidden information, which can increase the efficiency of the production processes.

In the context of the current technology direction, it is necessary to prepare solved tasks for their use in the Industry 4.0 environment using the principles of Industrial IoT. Industry 4.0 directly defines the approach to production, or a production process, such as the digitization of processes, products, and services and their interconnection. Suppose the Internet of Things (IoT) is defined as a system of interconnected computing, mechanical, and electronic devices with unique identifiers and the ability to transmit data over a network, without the need for human-human or human-computer interaction. In that case, the Industrial Internet of Things (IIoT) refers to interconnected sensors, instruments, and other devices networked with computers for industrial applications, including manufacturing and energy management. This connectivity enables acquiring, exchanging, and data analysis, potentially facilitating productivity and efficiency, as well as other technical and economic benefits. In a certain way, the IIoT represents a form of extension of the distributed control system, which enables a higher degree of automation using cloud computing to refine and optimize process control.

Trnava, Slovakia   Juraj Ďuďák
Žilina, Slovakia   Gabriel Gašpar
December 2022

Acknowledgments

Writing a book is harder than we thought and more rewarding than we could have ever imagined. Years of research and development are included in this work. We would like to thank our colleagues in the development team who supported us in our work with their expertise and constructive solutions to technical problems. Thanks to Ivan Sládek, who largely designed the hardware solutions, and Martin Skovajsa for technical support in the preparation of materials. Roman Budjač, Matej Fitoš, Matúš Nečas, and Jakub Perička played a significant role in the peer review of the final text. Friends, thank you. Considerable efforts were made by prof. Ružica Nikolić in translating and checking the translation of this book. Many thanks to the Research Centre of the University of Žilina. Specially prof. Branislav Hadzima for providing the conditions for the successful completion of this book.

Reviewers

Our great thanks also go to the peer reviewers who, in their demanding work, have found space for constructive criticism of our forthcoming book. Thanks to:

- *Prof. Dr.-Ing. habil. Peter Husár,* Technical University of Ilmenau, Institute of Biomedical Engineering and Informatics, Ilmenau, Germany
- *Prof. Ing. Dušan Maga, Ph.D.*, Czech Technical University in Prague, Faculty of Electrical Engineering, Prague, Czech Republic.
- *Prof. dr. sc. Neven Vrček*, University of Zagreb, Faculty of Organization and Informatics, Varaždin, Croatia.

Contents

1   **Communication Protocols**... 1
    1.1   Communication Standard EIA-485 1
    1.2   MODBUS/uBUS Serial Line Protocol............................... 4
    1.3   Design of the uBUS Application Protocol 10
        1.3.1   Secure Communication Layer.................................... 17
        1.3.2   Organizing the MultiSlave Modules 27
        1.3.3   Specification of the uBUS Protocol........................... 30
    1.4   One-Wire Communication Protocol.................................... 52
        1.4.1   1-Wire Application Protocol 58
        1.4.2   1-Wire Master... 60
        1.4.3   1-Wire Slave 63
        1.4.4   Dynamic Control of the 1-Wire Bus Status................... 71
    References 75

2   **Hardware Modules of the Sensory System** 77
    2.1   One-Wire Booster Module... 80
        2.1.1   Communication Interfaces and Microcontroller 82
        2.1.2   Modified Controller of the 1-wire Bus....................... 83
        2.1.3   Detection of a Short-Circuit on the 1-wire Bus and
                Current Consumption Monitoring 85
        2.1.4   The Bus Length Detection..................................... 86
        2.1.5   Improved Active 1-wire Bus Pull-up 89
        2.1.6   Mechanical Implementation of the 1-wire Bus Sensors 91
        2.1.7   Implementation of Sensors for Ambient
                Temperature Monitoring....................................... 91
        2.1.8   Unique Designs for 1-wire Temperature Sensors 92
    2.2   Measuring Module RRM 99
        2.2.1   The RF Communication Layer................................. 101

 2.3 Measuring Module THB 103
  2.3.1 Application Communication Interface of the
     AM2303 Sensor.. 106
  2.3.2 Implementation of Sensors for the THB Measuring Module . 107
 2.4 The Measuring TNZ Module 108
  2.4.1 Implementation of the Specific uBUS Protocol Commands.. 114
  2.4.2 Implementation of Sensors for the TNZ Measuring Module . 115
 2.5 Sensory Modules.. 117
  2.5.1 The OWS Sensory Module 117
  2.5.2 The RTM Sensory Module 128
 References 139

3 Software Modules of the Sensory System 141
 3.1 Basic Software Modules 142
  3.1.1 nSoric IS—Data Model of the Sensory System 142
  3.1.2 nSoric senlib—Shared Sensor System Library 145
 3.2 Server Modules.. 147
  3.2.1 nSoric API 148
  3.2.2 nSoric Serve.. 155
 3.3 Software Applications 157
  3.3.1 nSoric Merula... 158
  3.3.2 nSoric Aurela 160
  3.3.3 nSoric Cofig.. 165
 3.4 Sensor System Integration with the IoT Networks 168
  3.4.1 Measuring Modules and IoT 170
  3.4.2 Implementation of a Local Measurement Module 173
 Reference 174

4 Practical Use-Case of Proposed Measurement System 175
 4.1 Monitoring of Ecological Building Insulation 176
  4.1.1 Technical Solution 177
 4.2 Temperature Monitoring in Bulk Biological Materials................ 181
  4.2.1 The Hardware Part of the System 182
 4.3 Production Hall Temperature Monitoring............................ 184
  4.3.1 Technical Solution 184
  4.3.2 Software Solution... 185
 4.4 Monitoring of the Asphalt Road and Its Subgrade Freezing 192
  4.4.1 Experimental Installation 194
 4.5 Experimental Measurement of Asphalt Pavement Fatigue............ 196
  4.5.1 Deformation Measurement 199
 4.6 Authors' Protected Intellectual Property 200
  4.6.1 Sensor for Measuring Mechanical Deformation of
     Asphalt Pavement.. 201
  4.6.2 Universal Serial Bus Device for Measuring Physical
     Quantities 202

4.6.3   Sensor for Measuring the Temperature Profile of the
        Asphalt Pavement... 202
4.6.4   System for Wireless Measurement of Environmental
        Variables in Biologically Active Materials 204
4.6.5   Rod Probe for Temperature Measurement with an
        Adjustable Position of Installed Sensors 204
References 205

Glossary 209

Index... 211

Acronyms

ADC	Analog/Digital Converter
ADU	Application Data Unit
AES	Advanced Encryption Standard
API	Application Programming Interface
ASK	Amplitude-Shift Keying
AWG	American Wire Gauge
CLI	Command Line Interface
CRC	Cyclic Redundancy Check
CSV	Comma Separated Values
DAC	Digital/Analog Converter
DES	Data Encryption Standard
DCS	Distributed Control System
EC	Error Code
EEPROM	Electrically Erasable Programmable Read-only Memory
ESD	Electrostatic Discharge
FC	Function Code
GND	Ground
GPIO	General Purpose Input/Output
HDPE	High Density Polyethylene
HVAC	Heating Ventilation and Air-Conditioning
IDW	Inverse Distance Weighting
IIoT	Industrial Internet of Things
IoT	Internet of Things
I^2C	Inter-Integrated Circuit
JSON	JavaScript Object Notation
JWT	JSON Web Token
LDO	Low Dropout
LSB	Least Significant Byte
MAC	Medium Access Control
MDSS	Maintenance Decision Support Systems
MSB	Most Significant Byte

NWS	National Weather Service
OOK	On-Off Keying
OSI	Open Systems Interconnection Reference Model
PDU	Protocol Data Unit
PPR	Polypropylene Random Copolymer
PROM	Programmable Read-Only Memory
PUR	Polyurethane
RF	Radio-Frequency
ROM	Read Only Memory
RSA	Rivest–Shamir–Adleman Cryptosystem
RTD	Resistance Temperature Detector
RTU	Remote Terminal Unit
RWIS	Road Weather Information System
RWS	Road Weather Station
SPI	Serial Peripheral Interface
SR	Super Request
SRAM	Static Random Access Memory
SSL	Secure Sockets Layer
UART	A Universal Asynchronous Receiver-Transmitter
UML	Unified Modeling Language
WEP	Wired Equivalent Privacy
WPA	Wi-Fi Protected Access

Chapter 1
Communication Protocols

Where there is a will, there is a way. A communication way.

Standardized communication industrial interfaces are used in the development of measurement and sensor modules. In this chapter, we present a description of these interfaces to the extent necessary for their subsequent implementation into the firmware of the designed measurement and sensor modules. For control modules communication needs (in most cases, it is an industrial control computer) the serial communication using the EIA-485 standard is used, which allows for the half-duplex, multipoint connection. The uBUS [1] was chosen as the application communication protocol, which is a modification of the MODBUS protocol to expand the functionality of the hardware module that implements this protocol. In implementing individual sensors with the standard I^2C bus, 1-wire bus was also used.

1.1 Communication Standard EIA-485

EIA-485 (or RS-485) is a serial communication standard defined in 1983 by the EIA (Electric Industries Association). The current version is labeled TIA /EIA-485-A "Electrical Characteristics of Generators and Receivers for Use in Balanced Digital Multipoint Systems" from 1998 [2, 3].

In a digital multipoint system, the standard defines the characteristics of the generators and receivers. Details like signal strength, time, protocol, pin assignments, power supply voltages, and working temperature range are not mentioned. Two or more generators and one or more receivers make up a multipoint system. Since a generator is identical to a transmitter and multiple transmitters can share an electrical bus, EIA-485 is appropriate for multimaster systems [4].

The standard describes the characteristics of the generators and receivers used in a digital multipoint system. Signal quality, time, protocol, pin assignments, power supply voltages, and working temperature range are a few other details that are left

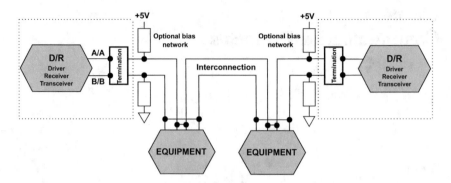

Fig. 1.1 Typical EIA-485 network

out. A multipoint system is made up of two or more generators and one or more receivers. Multimaster systems can use EIA-485 since the generator functions the same as a transmitter, and multiple transmitters can coexist on the same electrical bus.

Typically, an EIA-485 bus has two or more communication controllers, each powered by a different power source (Fig. 1.1). The controllers are daisy-chained together using at least one shielded or unshielded twisted-pair cable. The drivers, receivers, and transceivers connected to the network must adhere to the EIA-485 specification. Consequently, specifications are made for variables such as unit loads, output drive, short circuit current, and standard mode voltage. Under a standard mode voltage range of −7 to +12 V DC, a driver must be able to source at least 1.5 V differentially into 60 Ω (two 120 Ω terminators connected in parallel with 32 unit loads). No data rates are mentioned. Numerous devices that are compliant with the standard can operate at high (up to 50 Mbps) or low speeds (skew rate limited).

The sole layer that EIA-485 describes in terms of the Open Systems Interconnection Reference Model (OSI) is the physical layer.

When deploying EIA-485 networks, several crucial issues need to be taken into account, including termination, fail–safe bias, connectors, grounding, cabling, and repeaters [4].

- **Termination.** Reflections that could result in data mistakes are minimized by terminating a data cable with a value equal to its characteristic impedance. However, termination might not be required if the data rate is modest or the wires are short. Termination becomes increasingly important as the data rates rise. Since any bus device can transmit, a node amid the bus is likely to do so, necessitating the application of termination to both ends of the bus segment. Although twisted-pair cable impedance can be as low as 100 Ω, resistive terminators commonly have 120 to 130 Ω values. A 100 Ω terminating resistor is too low for the EIA-485 drivers. It is necessary to apply a value at a convenient point as close to the ends of the cable segment as possible that closely matches the cable impedance.

- **Fail–safe bias.** Individual devices share a common two-wire medium to send and receive data according to the multipoint EIA-485 standard. Due to the high likelihood of collisions caused by two transmitters operating simultaneously, a medium access control (MAC) technique is necessary. Since a bus arbitration method is a data link layer requirement rather than a physical layer requirement, the EIA-485 standard does not provide one.
- **Connectors.** EIA-485 standard does not cover connectors. Manufacturers or trade organizations must comply. There appear to be three common strategies based on practice. With a four-pin, six-position, or eight-position RJ-11 or RJ-45 connector, the conventional method offers plenty of signal and ground reference pins. The use of detachable open-style screw connectors is an alternative connecting strategy. A common connector used with the Profibus standard is a DB9 connector.
- **Grounding.** EIA-485 is a three-wire system. The requirement for a return path between the circuit grounds at each end of a connection for generators and receivers is explicitly stated in the standard. A physical connection in the cable linking each logic ground might serve as this return way, or the Earth (Fig 1.2) could serve as this return path by having each logic ground returned to it. The latter method can be utilized with a single-pair twisted cable. The third wire must have some resistance to the logic ground if it is to be used, according to the standard, in order to prevent circulating currents when alternative ground connections are offered for security. The resistor may be positioned between the logic ground and either the frame (connected to the earth ground) or the third connection. The standard cites $100 \, \Omega$ as an example for both scenarios.
- **Cabling.** The choice of a cable is one of the more critical choices to be made. It is mistakenly believed that any 24 AWG (American Wire Gauge) telephone wire will work when there is a wide variety of cables available. Several variables,

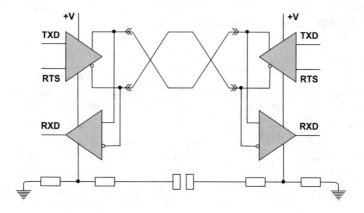

Fig. 1.2 Typical EIA-485 grounding

including data rate, signal encoding, and desired distance, influence the choice
of cable. The transmitted signal is weaker, and the signal's waveform is distorted
due to cables.

1.2   MODBUS/uBUS Serial Line Protocol

The uBUS application protocol is based on the MODBUS protocol. The physical
layer is taken from the MODBUS protocol or serial line specification [5]. From
the implementation into the ISO/OSI model point of view, individual layers can be
classified according to Table 1.1. The MODBUS serial line protocol is a master–
slave protocol. This protocol works on the second layer of the 7-layer ISO/OSI
reference model. The master–slave communication system has one node that issues
commands (master) to one of the slave nodes and then processes the responses.

According to the specification [5], the MODBUS serial line protocol should be
implemented according to the EIA/TIA-485 standard [2], which is also referred to
as RS-485. There are two versions of implementation: 2-wire implementation and
4-wire implementation. The MODBUS serial line protocol can also be implemented
on an EIA-232 [6] (RS-232) physical interface. In the standard MODBUS protocol,
all devices are connected (parallel connection) to a bus consisting of 3 wires. Two
wires are (2-wire configuration) a symmetrical 2-wire twisted-pair line through
which the data is transmitted in both directions. A typical speed is 9600 baud.
The MODBUS serial line protocol should implement a 2-wire electrical interface
according to the EIA/ITA-485 standard (Fig. 1.3). On such a 2-wire bus, only one
device can transmit at a time. Third, the common conductor of this bus must be
connected to all the connected devices.

The MODBUS serial line protocol connects one master device and several,
but no more than 247 slave devices to the bus. The master node always initiates
communication. The slave nodes never send data without receiving a request from
the master node. A master node initializes only one transaction at a time. The master
node sends messages to slave nodes in 2 modes:

Table 1.1 Protocol MODBUS—a reference model ISO/OSI

Layer	Model ISO/OSI	MODBUS
7	Application layer	MODBUS/uBUS application protocol
6	Presentation layer	–
5	Relation layer	–
4	Transport layer	–
3	Network layer	–
2	Link layer	MODBUS serial line protocol
1	Physical layer	EIA/TIA-485

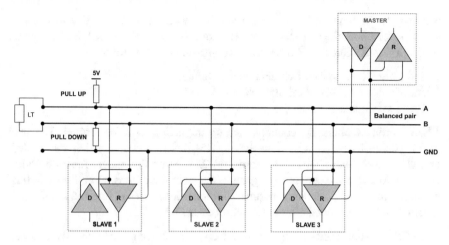

Fig. 1.3 Typical EIA-485 network

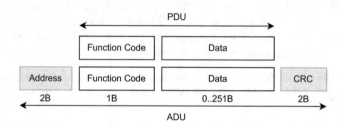

Fig. 1.4 MODBUS protocol—PDU and ADU units

1. Unicast mode. The master addresses a specific slave node. After receiving and processing the request, the addressed slave node responds to the master node. Each slave has a unique address (1 to 247), guaranteeing unique identification.
2. Broadcast mode. A master node can request all the slave nodes at once, and slave nodes do not respond to broadcast requests. Broadcast requests are used to send commands to all the nodes and all devices must accept a broadcast message with a data write function.

MODBUS Protocol Data Frame The MODBUS application protocol [7] defines a Protocol Data Unit (PDU) independent of the lower communication layers. Additional data fields are added to the PDU to implement the MODBUS protocol for a specific bus or network. The node that initiates the MODBUS serial line transaction creates an extended MODBUS frame ADU (Application Data Unit), where additional fields are added (Fig. 1.4).

Address field—according to the specification of the MODBUS serial line protocol [5], this field contains addresses of the slave nodes only. The size of this part is 1 byte. Acceptable slave node addresses are 1 to 247. The master node addresses the slave node by placing the address in the Address field. When the slave node responds to the request, it puts its address in the Address field so that the master knows from whom the response came. The address space of the MODBUS protocol is 255 different addresses. Addresses are divided into three groups:

1. Indexbroadcastbroadcast communication address: 0
2. Individual addresses of the slave nodes: 1–247
3. Reserved addresses: 248–255

The Function Codefield—contains data on the action to be performed. The size of this part is 1 byte, which allows having 256 different codes, which are divided as follows: acceptable codes are from 1 to 255. From this range, function codes from 128 to 255 are reserved for signaling the error conditions. Function code 0 is not accessible. Some functions require additional information, referred to as "subfunction code" and added after the function code (Table 1.2).

Data field—when transmitting a frame from the master node to the slave node, the Data section contains parameters of a function the slave should perform. In the opposite direction, i.e., from the slave node to the master, the Data section contains values measured by the slave node. The Data section can contain various items such as discrete data, register addresses, and the number of data to be processed. The Data section is optional. The maximum length of the Data part is 252 bytes (Table 1.2).

Check field—the error check field results from the CRC sum calculated from the message's content. The MODBUS protocol uses the following polynomial (Eq. 1.1):

$$G(x) = x^{16} + x^{15} + x^2 + 1 \tag{1.1}$$

If no error occurred during the data transmission or processing, the response of the slave node contains the data requested by the master node. The Address and Function Code parts remain unchanged. If an error occurred during the processing of the request, then in the Function Code section of the message, there is an error code with which the master node can determine further actions. That error code is obtained from the original code, the most significant bit of which is changed to a value logical 1, or in other words, the value 128 (or 0x80) is added to the function code. An error message consists of a function error code and an error code (*Exception Code*). The slave node response types can be characterized as follows:

- Response without an error. Function code in the response = function code in the request.

Table 1.2 MODBUS protocol—PDU and ADU units [5]

Slave address	Function code	Data	CRC
1 byte	1 byte	0..252 bytes	2 bytes

Table 1.3 MODBUS protocol error codes [5]

Exception code	Error name	Description
01	ILLEGAL FUNCTION	The slave node cannot process the requested function
02	ILLEGAL DATA ADDRESS	Invalid address of slave node registers. The required registers do not exist
03	ILLEGAL DATA VALUE	An incorrect data structure in the Data section
04	SLAVE DEVICE FAILURE	An error of the slave node
05	ACKNOWLEDGE	Confirmation of the frame acceptance in the case the request processing will take longer than waiting for a response
06	SLAVE DEVICE BUSY	Response to the master node if the slave node is busy
08	MEMORY PARITY ERROR	Used in the case of working with files to report the file error
0A	GATEWAY PATH UNAVAILABLE	It is related to the use of the Internet/intranet gateways. Most of the time, the error occurs when the gateway is overloaded or misconfigured
0B	GATEWAY TARGET DEVICE FAILED TO RESPOND	It is related to the use of the Internet/intranet gateways. No response was received from the target device

- Response with an error occurrence. Function code in the response = function code in the request + 0x80. The error code *Exception Code* is determined according to the type of error that occurred (Table 1.3).

Status Diagrams of the MODBUS Protocol The link layer of the MODBUS protocol includes two sublayers: the master–slave protocol and the transmission mode (ASCII or RTU). The following state diagrams represent communication between the master and slave devices in the MODBUS protocol, regardless of the used transmission mode. The notation used for edges in the state diagrams is the standard UML notation: "trigger [control]/action." It has the following meaning: a trigger is an action that causes a state change; the state change occurs only if the conditions defined in the control section are met, and the action represents the event that will be executed. The following state diagram explains the behavior of the master device [5].

Notes on the state diagram in Fig. 1.5:

- The status "Idle" means the master has no request to send. It is the initial state after the bus is turned on. The request can only be sent in the Idle state. After sending a request, the master changes its state and cannot send another request.
- If the slave node has sent a request of the unicast type, the master changes its state to Waiting for a response and the timer $t1$ is initiated (time limit for receiving a response—Fig. 1.6). This timer prevents the master from remaining

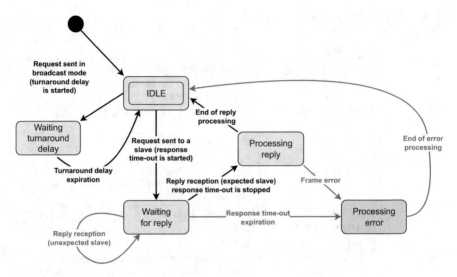

Fig. 1.5 Modbus master state diagram [5]

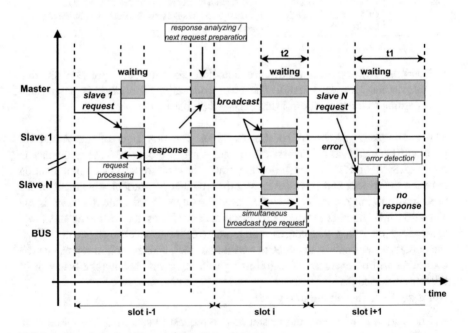

Fig. 1.6 Communication diagram in Modbus protocol

in the Waiting for Response state. The value of timer t1 depends on the running application.

- When a response is received, the master checks the message before processing the data. The check may detect, for example, a response from an unexpected node or an error in the received frame. If a message was received from an unexpected slave node, the timer t1 remains active. A retransmission may be required if an error was detected in the frame.
- If no response is received at time t1, an error is generated. The master changes its status to Idle. In this state, it can request the message to be resent. The maximum number of repeated request forwarding depends on the setting of the master node.
- If a broadcast request was sent to the bus, no response is expected from the slave nodes. After sending such a request, the master waits for time t2 (time limit for processing the request—Fig. 1.6), during which the slave nodes process the request. This time limit prevents sending a new message while the nodes can still process a previous request. The master changes its state to Waiting t2 for time t2. After that time, it will return to the Idle state.
- In the unicast communication mode, the time t1 must be set long enough for any slave node to process the request and return the response to the master node. In broadcast communication, the time t2 must be long enough for all the nodes to process the received message and to be able to receive the next message. Therefore, time t2 should be shorter than time t1. The value of t1 is usually from 1s to several seconds at a speed of 9600 bit/s. The value of t2 is from 100 ms to 200 ms.

Status diagram of the functionality of the slave device is in Fig. 1.7. There is an explanation of this state diagram.

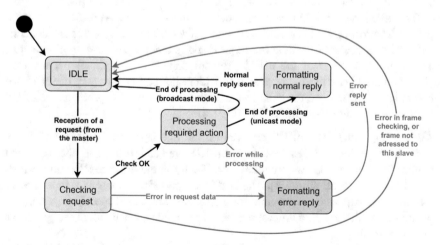

Fig. 1.7 Modbus slave state diagram [5]

- Status "Idle" means no pending request and is the initial state.
- After receiving the request, the slave node checks the received packet before performing the requested action. Various errors may be detected during the check, for example, a formatting error, an unknown action, and others. If an error is detected, a response must be sent to the master node.
- The slave node must send a response to the master node if the requested action was performed. This operation only applies to the unicast communication mode.
- If the slave node detects an error in the received frame, no response is sent to the master node.

1.3   Design of the uBUS Application Protocol

This section presents a proposal for extending the MODBUS protocol, called uBUS. This new specification is based on the original MODBUS serial line protocol specification. The objective of the proposed uBUS specification is to increase the applicability of the MODBUS protocol in real industrial applications. The first modification is the modification of the slave device, where by the slave device, we mean several functional hardware components that can communicate with one another—Fig. 1.9. These components are considered as one slave device. Such a device (or multi-device) solution is labeled MultiSlave [8]. The second modification of the slave node is concerned with its functionality. The proposal was named SuperSlave. In Fig. 1.12 is a timing diagram of communication when the master node's task exchange function is activated. Functionality is added to the slave device when it takes over the role of the master device for a specified time [8].

The uBUS protocol defines a basic communication frame—Protocol Data Unit (PDU). The format of it is identical to the MODBUS protocol. This frame contains only the data needed to handle the request and is independent of the other layers. The application layer of the uBUS protocol defines the Application Data Unit (ADU), which adds the fields related to addressing and consistency check of the communication frame to the PDU frame. Unlike the MODBUS protocol, the uBUS protocol uses 2 bytes for addressing (Fig. 1.4).

Addressing Rules The uBUS protocol on the line layer uses all the MODBUS serial line specification features. In the uBUS application layer, a change of the application framework was proposed, specifically in the "Address" section, which now has a size of 2B (Fig. 1.4). The "Address" part size change is also related to the change or expansion of communication modes. The uBUS specification defines, in addition to the standard unicast and broadcast modes, a new mode designated as group broadcast.

In communication, there are three ways of addressing, which differ in how the end node or nodes are addressed. Here are the types of communication for the uBUS protocol:

Fig. 1.8 uBUS addressing [1]

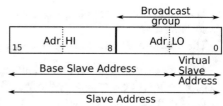

Fig. 1.9 MultiSlave extension [8]

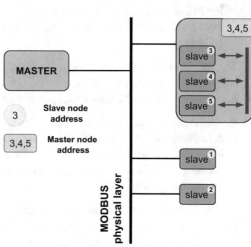

1. Unicast—master addresses a specific slave. Once the request has been received, the slave always responds. The uBus protocol recognizes the standard slave and the virtual slave. The standard slave is a hardware implementation of a slave whose addresses can be from 16 (0x010) to 4095 (0xFFF) or 4079 different addresses. Each standard slave may contain 1 to 16 virtual slaves whose addresses can be from 0x0 to 0xF. The address of the slave is, therefore, a 2-byte value from which the first 12 bits represent a standard slave address and the last 4 bits represent a virtual slave address (Fig. 1.8).
2. Broadcast—the master sends a message to all the connected slaves. Slaves do not answer the request. In the broadcast mode, the Adr_HI and Adr_LO bytes are set to 0.
3. Group Broadcast—the master sends a message to a slave group. Any slave does not answer the request. The upper byte of the Adr_HI address is 0x0, and the lower byte defines the broadcast group. Addr_LO value can be from 1 to 255.

MultiSlave The proposed MultiSlave extension represents several simple slave devices, which are physically designed as a single device. Since there are multiple slave devices, each slave device must have its address (Fig. 1.9). From the Multi-Slave node functionality point of view, the individual modules are connected, for example, via an internal bus.

The following modifications are proposed for the MultiSlave node:

1. The MultiSlave node consists of a set of slave nodes as defined by the MODBUS specification.
2. The MultiSlave node is assigned a set of addresses.

    a. A message with an address belonging to the set of addresses of the MultiSlave node must be accepted by the MultiSlave node.
    b. The number of addresses of the MultiSlave node is more than 1. Both the hardware parts of the MultiSlave node and the virtual hardware blocks must have an address assigned (the function of the virtual block is provided by the software).

3. The modules of the MultiSlave node form a connected subsystem capable of communicating with each other.
4. Each submodule works with one input or output interface/sensor.

    a. A submodule can represent one sensor.
    b. A submodule can represent an interface to another subsystem (communication bus bridge)—for example, an expansion bus to which other sensors are connected. In this case, such a submodule manages all the connected sensors on this external bus.

5. The MultiSlave node responds with a uBUS protocol data frame. The address field in this response has the value of the address in the request.
6. The status diagram of the MultiSlave node is shown in Fig. 1.10.

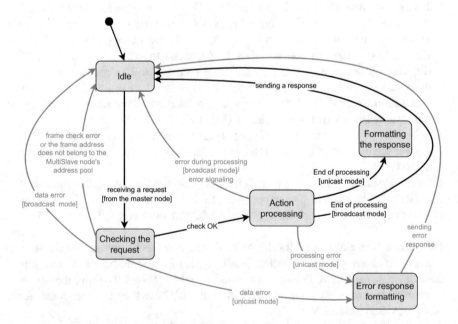

Fig. 1.10 Proposed state diagram for MultiSlave node [8]

The state diagram of the MultiSlave node is based on the state diagram of the slave node in Fig. 1.7. The proposed state diagram is supplemented with an extended check of the address in the received frame in case it is a MultiSlave type node. The received frame is accepted if the address in the received frame belongs to the set of addresses of the MultiSlave node.

SuperSlave The design of this extension was to enable slave/master communication in specific cases. The need for such communication may appear when transmitting certain information (for example, real-time information) to all the slave nodes. However, this information is available to only one slave device. The standard solution in the MODBUS protocol is that the master asks the slave for the particular information, and then the master sends it as a broadcast message. The solution in the uBUS protocol (SuperSlave extension) is that the master instructs the slave node to send the information to the broadcast address. The principle of master/slave communication functionality guarantees that there will be no conflict on the bus. This is achieved by allowing the slave devices to transmit on the bus only when requested by the master device. The principle of communication will not be violated even if the master instructs some slave device to send a single broadcast message for a precisely defined time. The proposed modification allows for adjusting the slave node functionality, to which the master device can delegate the right to send broadcast messages for a precisely defined time. With this modification, the solution for the previous situation will be as follows: the master device sends a request to the slave node to send the time value to all the other slave nodes. The slave device sends a broadcast message with the required data. After the defined time, the given slave device loses the right to send data and the master device can process further requests.

The following modifications are proposed for the SuperSlave node:

1. The SuperSlave node must be backward compatible, meaning that it must contain the full functionality of a slave node.
2. After receiving a special request, the SuperSlave node processes the request and sends the processing results to the bus as a broadcast message.

   a. The SuperSlave node has the right to send a broadcast message to the bus only for a precisely defined time (tw). That time is defined directly in the received message.
   b. After that time, the SuperSlave node must not start broadcasting the message.
   c. A SuperSlave node can send only one broadcast message at a time.

3. The SuperSlave node only responds to a broadcast transmission request with a broadcast message.
4. A SuperSlave node cannot send unicast messages in response to a broadcast request.

In the original specification of the MODBUS serial line protocol, the slave node must respond to each unicast request with a message that contains the address of the addressed slave node. In the case of success, the function code in the response is the same as in the request. When receiving a super request in SuperSlave modification,

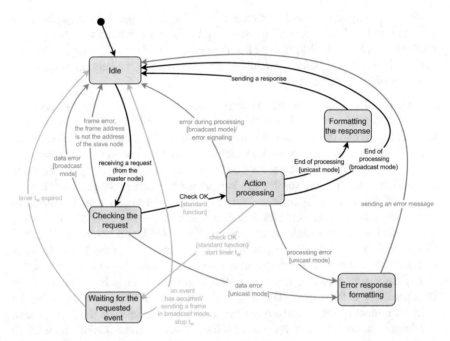

Fig. 1.11 Proposed state diagram for SuperSlave node [8]

we remove the obligation to respond with such a message since the slave node would first have to respond to the master node and then send a broadcast message. However, such a solution does not bring the expected improvements in the protocol properties. If a slave node is requested to transmit a broadcast type, then this broadcast type transmission is taken as a response to the request.

The design of the state diagram for the SuperSlave node is shown in Fig. 1.11. It differs from the original diagram by adding the "Waiting for the requested event" state, which is reached only if the request check went without problems and is a "super request." The SuperSlave device remains in this state for a maximum period defined in point 2.a of the definition of the SuperSlave node. If the SuperSlave node manages to handle the request in the received message by this time, it sends a broadcast message as its response. If the SuperSlave node fails to handle the request by the time t_w, it must not send any response. In both cases, the status will subsequently change from "Waiting for the required event" to "Idle."

In the proposed modification, some changes also concern the master device. Therefore, we define the following rules for the master device in addition to the standard specification of the MODBUS master protocol.

1. The master node must be backward compatible with the MODBUS serial line protocol specification.
2. Special requests for a SuperSlave node can only be sent as a unicast message.

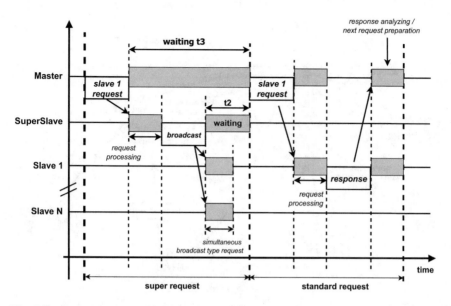

Fig. 1.12 Timing diagram of SuperSlave extension communication [8]

3. After sending a special request to the SuperSlave node, it will wait for a precisely
   defined time (t_3) to complete the processing. See Fig. 1.12 or relationship 1.2.

   a. In response to such a request, the master node must accept only a broadcast
      message.
   b. If time t_3 has expired and no response has been received from the Super-
      Slave node, the SuperMaster device starts processing this error condition.
      Depending on the nature of the application, the request may be resubmitted or
      dropped.

In Fig. 1.13 are indicated suggested changes to the state diagram of the master
node. Changes are displayed in green (if communication was without errors) and
red (processing of error status).

   The following relationship applies to time t_3 (according to the diagram in
Fig. 1.12:

$$t_3 = t_w + t_{BC} + t_2 \tag{1.2}$$

where t_w is the maximum time the SuperSlave node has to process the request. Time
t_{BC} is the time of sending a broadcast message to the bus. Since the length of the
message is variable, this time is not constant at a given transmission rate. The time
t_2 is when the master node waits before returning to the idle state after sending a
broadcast message. The time t_3 depends on the running application. The time t_w
value is determined by the master node or by the user application according to the

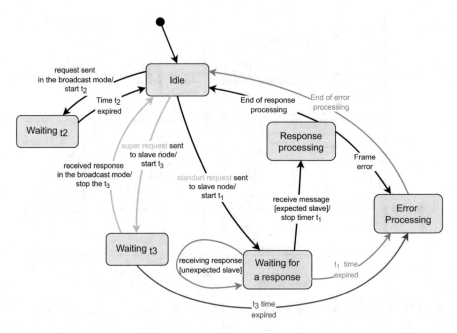

Fig. 1.13 Proposed state diagram for Master—SuperMaster node [8]

nature of the super request. We can derive the value of t_{BC} for a given transfer rate fp as

$$t_{BC} = \frac{\text{frame\_length}}{\text{fp} \cdot \text{bit\_per\_baud}} \tag{1.3}$$

which, with a maximum data frame length of 256 bytes, a baud rate of 115,200 baud, and a baud rate of 9 (8 bits, 1 stop bit, no parity), is a value of about 20 ms. The typical length of a broadcast message will be shorter than the maximum length. If we consider *frame_length* = 32 B, then t_{BC} = 2.5 ms. We can consider the times t_w and t_2 to be the same. The resulting time t_3, therefore, depends mainly on the time $t_2 + t_w$, which is the maximum waiting time for the broadcast request processing. Depending on the nature of the application, it can range from units of milliseconds to hundreds of milliseconds.

Format of SuperSlave Node Requests The request formats, sent by the Super-Slave node and the format of the response, are shown in Table 1.4:

- Request frame:

  - SR—Super request code. Codes 0x55—0x5F are reserved for super requests.
  - t_3—the maximum time for processing the request and ending the communication in milliseconds.

Table 1.4 Specification of the data frame for the SuperSlave node request

Action/byte	0—Adr	1—Adr	2—FC	3 .. 4	5	6
Request to SuperSlave	0xNN	0xNn	SR	t_3	FC	DATA
Response SuperSlave broadcast	0	0	FC	DATA	CRC1	CRC2

- FC—function code for the broadcast request to be sent by the SuperSlave node.
- DATA—additional data to the request. They are not mandatory.

• Response of the SuperSlave node—broadcast request:

- The function code and data are taken from the original request.

1.3.1 Secure Communication Layer

According to the nature of devices, industrial buses are divided into sensor buses (Sensorbus), buses connecting devices (Devicebus), and own industrial buses (Fieldbus). The proposed sensor system belongs to the Sensorbus category. Since there is no communication security method for the sensor networks of the Sensorbus category, this layer was added to the proposed system.

The main task of the sensor network is to ensure the reliable transfer of measured data from the sensors to the control unit, where they can be further processed. Due to the nature of data, it is necessary for those to be true or to guarantee their authenticity. A standard part of communication is the checksums, which carry information about the communication frame's consistency. A potentially dangerous situation can appear when an element is added to the sensor network that purposefully sends incorrect data. Due to the nature of the sensor network, this can be a severe security issue, which can even result in material damage. A proven way to prevent such incidents is to secure communication with appropriate encryption. When implementing secure communication, one can consider two independent levels:

1. Ensuring communication at the lowest level of communication, i.e., from sensors to the measuring module and from the measuring module to the measuring station
2. Provision of data transfer from the measuring station to the application server

Generally known and proven methods, such as using the SSL layer, are available to ensure data transfer from the measuring station to the application server. In this part, we are mainly focused on ensuring communication at the lowest level—with measurement modules to which the sensors are connected directly or via the local 1-wire bus. These blocks are connected to the measuring station via the RS-485 serial communication bus.

1.3.1.1 Communication Layer Implementation[1]

Secure communication means encrypted communications frames at the sensor level (the lowest level) for the needs of measuring systems [9]. The algorithm (Fig. 1.14) for processing the uBUS protocol was designed to solve the protocol implementation on slaves. The algorithm assumes that the received frame has the correct APU and PDU format shown in Fig. 1.4. When processing the frame, the CRC16 is checked first. In the case of a disagreement, the slave does not respond and must be ready to retrieve the next frame. If the CRC16 is valid, the algorithm determines the communication type. No response is sent after the request processing for Broadcast and Group Broadcast modes. In the Unicast mode, the value of the address is first checked. Since the slave can contain multiple addresses (MultiSlave extension), the algorithm checks if the slave is a part of the current MultiSlave. The next step is detecting the communication encryption usage. Slaves have information about the encryption obligations recorded in their configuration settings. When attempting to proceed with unencrypted communication, the slave with enforced encryption sends an error response informing about the obligatory encryption. Details on the use of encryption are described in the following section.

MultiSlave consists of multiple slaves. From a communications point of view, one can divide the supported features into features of MultiSlave as a whole and specific slave-related features. Features related to MultiSlave include its name, firmware, hardware information, identification number, number of slaves, and encrypted communication support information. Each slave (such as a sensor or an actuator) must implement slave-specific features: slave identifier, slave type (measured physical quantity, physical units), and limit values (minimum and maximum). In addition to this information, it must support measurement-related functions: triggering the measurement and sending the measured value.

Communication Security Implementation In the case of non-secured communication, the attacker can quite easily take over the communication server role and gain full control over all the subordinate devices. Then it can set new values and parameters, even restart [10]. Securing communication must meet the following conditions:

1. Encryption must be strong enough to make immediate decryption (without knowledge of the decryption key) impossible.
2. Encryption must be effortless to implement at the slaves with small computing power.
3. Communications framework format should not be modified.
4. It must be easily identifiable, whether it is an encrypted or an unencrypted frame.

---

[1] This section describes the authors' solution, which was in the past partially published in IEEE Sensors Journal—[1].

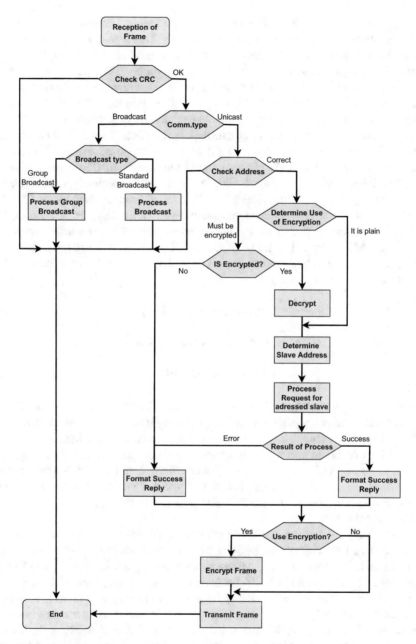

Fig. 1.14 Algorithm of the uBUS request processing [1]

We suggest the following solutions to achieve the required conditions: Ad (1) and (2)—the RSA asymmetric encryption will we used. This type of encryption ensures ease of implementation at the slave. In addition, it provides a strong cipher on the principle of a pair of keys—encryption and decryption. The force is given by a cipher key size (32–1024 bits). The length of the key, however, is limited by the slave hardware with implemented encryption. The "floating code" mechanism is implemented to enhance security. It will be used by the encryption key, valid for a specific time only (in the order of minutes). After that time, a set of new encryption keys will have to be generated. Ad (3) and (4)—proposed extension must be backward compatible with the plain communication in the uBUS protocol. It must be identifiable when the communication is encrypted (respectively, unencrypted). The format of the part "Error Check" of the communications framework ADU is used to identify the type of communication type. The byte order is used in the CRC16 checksum to determine the encrypted and unencrypted communications. Let the FRAME be the date frame of the uBUS protocol, consisting of addresses (2B), function code (1B), and data portion (0–252B). The CRC16 checksum can be written as

$$CRC = CRC16(FRAME) \tag{1.4}$$

The length of the CRC is 2B. Thus, one can write this as

$$CRC = CRC\_hi \; CRC\_lo \tag{1.5}$$

where the CRC_hi is the first byte of the CRC and the CRC_lo is the second byte of the CRC. In unencrypted communication, the byte sequence in the checksum is CRC_hi, CRC_lo. In the case of encrypted communication, each byte is cyclically shifted by one bit to the right. The resulting check product can be labeled as (CRC_hi >>> 1), (CRC_lo >>> 1), where the operator >>> i is a cyclic shift to the right by i bits. The algorithm in Fig. 1.15 was designed for the communication frame consistency check. The functionality principle of this algorithm is described in the following text.

When using encryption, the ADU communication packet format must not be changed (Fig. 1.4). This packet includes the slave address, PDU part, and CRC16. In order to meet this condition, communication encryption is only applied to the PDU part. The address and CRC16 fields remain unencrypted, even with encrypted communication. The encryption detection algorithm in Fig. 1.15 input is an ADU data packet. The output is two flags: Status and Secure. The Status flag informs about the validity of the ADU frame, and the Secure flag informs about the use of encryption in the accepted PDU frame. Initially, both flags were negative. If the first CRC check succeeds, the communication is considered valid (Status = OK), and the encryption is not used (Secure = False). Otherwise, it follows a cyclic CRC byte shift by one bit to the right, followed by the second CRC check. If it succeeds, the communication is encrypted (Secure = True, Status = OK). Any other result means

Fig. 1.15 Algorithm of
communication frame
consistency check [1]

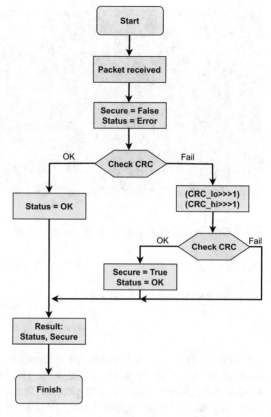

Table 1.5 Result of the CRC
check algorithm [1]

Status	Secure	Result—state
Error	False	The communication frame is valid
Error	True	Unauthorized state
Ok	False	PDU frame is not encrypted
Ok	True	PDU frame is encrypted

that the communication frame is considered to be damaged. Table 1.5 lists allowed
combinations of the algorithm states.

For reasons of implementation simplicity and security level of the used encryption, we selected the RSA algorithm. This algorithm uses two encryption keys:
public and private. The public key is used for encryption, and the private key is
for decryption. The RSA algorithm is defined as follows:

Algorithm RSA

1. Choose two different large random prime numbers p and q
2. Calculate $n = pq$; n is the modulus for the public key and the private keys
3. Calculate the totient: $(n) = (p1)(q1)$
4. Choose an integer e such that $1 < e < (n)$ and e is coprime to (n), i.e., e and (n) share no factors other than 1; $gcd(e, (n)) = 1$; e is released as the public key exponent
5. Compute d to satisfy the congruence relation de1(mod(n)), i.e., de= 1 + $k(n)$ for some integer k. d is kept as the private key exponent.

The following relation applies to the encryption of source x data:

$$y = x^e mod\ n \qquad (1.6)$$

while for decrypting the data, y applies

$$x = y^d mod\ n \qquad (1.7)$$

To correctly implement the RSA algorithm, generating two pairs of keys, one pair for each communicating party, is necessary. The first step is the exchange of public keys between the communicating parties. The principle of initializing encrypted communication and encryption itself is illustrated in Fig. 1.16.

Initialization of encrypted communication between the master and a slave, namely MultiSlave, runs in unencrypted form. Master generates its private $(n1, d1)$ and public $(n1, e1)$ keys. The public key is then sent to a slave that stores this key and generates its public $(n2, e2)$ and private $(n2, d2)$ keys. Slave sends its public key in the response. Further communication must be in encrypted form already.

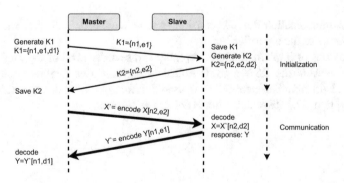

Fig. 1.16 Keys exchanging [1]

The master encrypts the message being sent with the public key $(n2, e2)$. The slave decrypts the message with its private key $(n2, d2)$ and encrypts the response with the public key $(n1, e1)$. The master then decrypts the message with the private key $(n1, d1)$. If the MultiSlave supports encrypted communication, the following use rules apply:

- Communication concerning MultiSlave information may be unencrypted. Functions to be used in an unencrypted form include, in particular, information on encryption support, initialization of the secure layer, and exchange of encryption keys.
- Communication related to a particular slave must always be encrypted. When trying to communicate unencrypted, the slave responds with an error state.
- Due to the nature of the unencrypted, only unicast communication will be encrypted.

The communication must be clear and unencrypted if the encryption is not supported. Otherwise, the MultiSlave will respond with an error code.

Implementation of the RSA Algorithm for STM32 Microcontrollers When choosing the size of the encryption keys, the performance options of the microcontroller and its instruction set must be considered. Cortex-M0 [11] and Cortex-M0+ [12] microcontrollers (STM32F0x, STM32L0x) have a significantly reduced instruction set.

(A) Minimalistic Implementation

Another requirement for implementing the encryption algorithms was the ability to implement this algorithm also on microcontrollers that do not have the hardware support for encryption. Only the microcontrollers from the "high performance" category have hardware encryption support. However, no high performance is required due to the complexity of the encryption (Eq. 1.6) and decryption (Eq. 1.7) rules. When implementing the RSA encryption algorithms, the following assumptions were made:

- An integer arithmetic with a 32-bit representation will be used.
- The encryption key size will be 32-bit.

From the practical point of view, the 32-bit key size is insufficient. However, this chapter focuses on the very principle of adding a secure layer to the communication protocol without changing the communication frame format. The algorithm described in Fig. 1.15 was designed to detect the use of the encryption algorithm in communication frames. In the implementation, a strategy of the minimal length of the communication framework was selected.

A test was implemented on six different microcontrollers of the STM32 family to compare the applicability of this minimal implementation of the RSA encryption for embedded solutions, a. Specifically, they were STM32F030 (ARM Cortex M0),

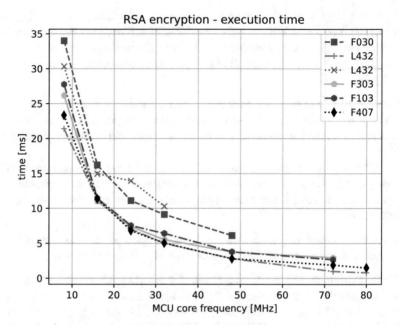

Fig. 1.17 RSA encryption algorithm—measured time of execution [1]

STM32L031 (ARM Cortex M0+), STM32F103 (ARM Cortex M3), STM32F303, STM32L432, STM32F407 (ARM Cortex M4). The test conditions were as follows: the length of the encrypted message was 12 bytes and the length of the encryption key was 32 bits. The graph in Fig. 1.17 shows the results at different set frequencies of the microcontroller core—from 8 MHz to 80 MHz. The minimum frequency at which the RSA algorithm can be used is 32 MHz, as seen in the graph.

(B) Basic Implementation

In the second version of the secure communication implementation, an encryption key length of 128 bits was calculated. The length of 8B for the encryption key is a compromise between the calculation speed of the RSA algorithm and the security level of the used cipher reliability. Since a key with a length of 8B is used, the minimum length of the communication packet is 12B (8B encrypted packet + 2B address + 2B CRC).

(C) Advanced Implementation

In this version of the secure communication implementation, the encryption key has a length of 256 bits. The length used is less than the minimum key length of

Table 1.6 Time complexity of the RSA algorithm with 512-bit key

MCU	STM32L0	STM32F0	STM32F4
Run time of RSA	42.3 s	22.3 s	3.6 s

2048 bits, recommended according to the NIST (National Institute of Standards and Technology) from 2015 [13]. However, this recommendation concerns the security of communication on the Internet. These sensory networks are where most short messages are transmitted. As the encryption key length increases, the minimum message length increases. If we have an average length of an unencrypted message of 8 bytes and for encryption with a key length of 256 bits, then the encrypted message will have a length of at least 32 bytes (+ 4B address and CRC).

When using an encryption key with a longer length (e.g., 512 bits and more), the difficulty of calculating the encryption/decryption algorithm becomes apparent. During the tests on real STM32 microcontrollers, specifically STM32L0, STM32F0, and STM32F4, with the same core frequency set $f = 32$ MHz, where a 16-byte long message was encrypted, the results were as shown in Table 1.6.

The results of the time consumption of the RSA algorithm on individual microcontrollers set to a speed of 32 MHz show why the RSA algorithm is used only to create the encryption keys for other faster encryption algorithms. Encrypting and subsequent decryption of a 16 Byte message, using the RSA algorithm with a 64 byte key, takes up to 42.3 s on the STM32L0 microcontroller, which is unusable for real applications due to the time requirement. For comparison, tests were performed with the same conditions (encrypted message length, key length, and microcontroller frequency) as for the RSA algorithm, using other known encryption algorithms: DES, TripleDES, AES, and HASH.

The graph in Fig. 1.18 shows the duration times of individual algorithms in milliseconds on all three microcontrollers at a set speed of 32 MHz, a key length of 256 bits, and a message length of 16B. As expected, the algorithms run the fastest on the STM32F4 microcontroller and the slowest on the STM32L0 microcontroller, which due to consumption uses only a two-stage pipeline. Furthermore, from the graph, the AES was the fastest encryption algorithm on all the microcontrollers. The slowest was the RSA encryption algorithm, whose duration is up to 5000 times longer than the second slowest TripleDES algorithm, which means that it is unusable for real-time communication [14].

Note that these algorithms (DES, TripleDES, AES, and HASH) were implemented using the STM32CRYPTOLIB library, distributed by the microcontroller manufacturer—ST Microelectronic Inc.

Padding Scheme An essential part of the RSA algorithm implementation is the method of selecting the blocks of data to which to apply relations 1.6 and 1.7. The input of the algorithm is the PDU frame (Fig. 1.4). The encryption and decryption key is a 32-bit, 128-bit, or 256-bit number. The input to the algorithm can only be a value smaller than n (part of the public and private key). The following assumptions

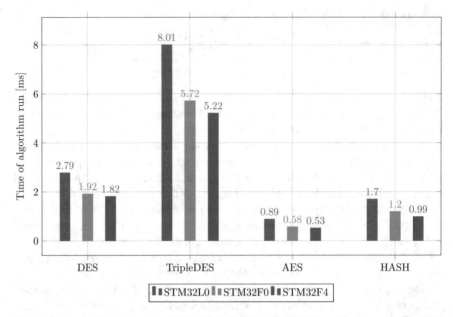

Fig. 1.18 Graph of duration times of individual algorithms at 32 MHz on all microcontrollers [14]

must be met to fulfill the condition that the encrypted value must be smaller than n:

1. The value of n must be greater than 0xFFFFFF (or 2^{24}) for a 32-bit key, 2^{120} for a 128-bit key, and 2^{248} for a 256-bit key.
2. The RSA algorithm is applied to the array of bytes, which will be divided into three bytes, fulfilling the condition for the input data range.

The algorithm of the communication frame distribution and its encryption is

Padding Algorithm

1. The communication frame is of the format (Fig. 1.4); for the needs of the RSA algorithm, we will work only with the PDU—Function Code and Data part.
2. Let the encryption key have a length of N bits. Let B denote the number of key bytes: $B = N/8$.
3. We will add zeros to the PDU from the right, so their number would be divisible by $(B - 1)$ without the remainder.
4. The number of these $(B - 1)$ byte blocks is denoted as NB.

(continued)

5. We apply the N-bit RSA cipher to the PDU block as follows:

   a. $Y = RSA(Xi)$ where Xi is a group of $(B - 1)$ bytes, s
   b. the result will be N Byte values, each of B bytes size.

6. We will add the encrypted Y data to the communication frame as a new part of the PDU.
7. Calculate the CRC16, cyclically shift it by 1 bit to the right and add it to the ADU communication frame.

The process of decrypting the frame is the same, except that the encrypted PDU communication frame is divided into 4 bytes, and decrypting result will be 3-byte values.

As shown by the results of measuring the time consumption of the RSA algorithm (Fig. 1.17 and Table 1.6), the use of a more secure encryption key (with a length of 512 bits) is unrealistic. The encryption time is acceptable when using an encryption key with a shorter length (32–128 bit) and a minimum microcontroller clock frequency of $f = 32$ MHz. The resulting implementation of the measurement module regarding the use of secure communication is as follows:

- The RSA algorithm was used with a key length of 32 bits (128 bits for the "Basic implementation" version). The measurement module contains a set of encryption keys that it uses during communication; the measuring station always generates its keys.
- Before starting communication between the measuring module and the measuring station, it is necessary to exchange public keys (the RSA algorithm is asymmetric),
- During communication, the measuring station can initiate the exchange of encryption keys at any time.

During the implementation of the secure layer, emphasis was placed on preserving the current functionality. Therefore, unencrypted communication is still supported. The measurement module decides whether secure communication will be used. At the start of communication, the measuring station addresses the measuring module and informs whether encrypted communication is supported. In the case of a positive answer, public keys are exchanged. When attempting unencrypted communication, the measuring module responds with an error message.

1.3.2 Organizing the MultiSlave Modules

The multislave module is a hardware module containing control, sensor, and communication parts. The sensor part depends on the specific implementation;

Fig. 1.19 Memory map for MultiSlave device

	0	1	2	3	4	5	6	7	8	9	10	11	12	14	14	15
0	ADR		SLAVES		SW VERSION				HW VERSION				N/A	DATE		
1	Multislave ID								-	-	CIPH	1W	FAMILY			
2	Multislave name								FW VERSION				-	-	-	-
3	-	-	-	-	-	-	-	-	-	-	-	-	-	-	-	-

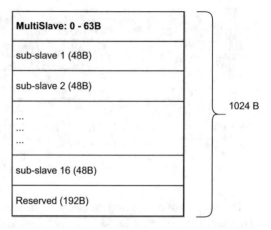

Fig. 1.20 MultiSlave memory map configuration

the communication part must respect the communication method described in the MODBUS/uBUS serial line chapter. Each multislave device must implement a way to store its configuration in EEPROM memory. The specific implementation is not defined. It is possible to use the internal FLASH memory of the microcontroller or external memory. The size of this memory must be 1kB. This memory is organized for setting the configuration of the measuring module as a whole (multislave) and the configuration of individual sub-slave modules. Figure 1.19 shows the basic distribution of the memory capacity of the configuration between individual parts of the solution. The memory is organized into lines of size 16B, so it contains a total of 64 lines.

The configuration of the MultiSlave module contains the following information (Fig. 1.20):

- ADR—module address, address: 0, length: 2B
- SLAVES—a list of implemented slave submodules. Address: 2, length: 2B. The format is described in the uBUS protocol command specification: Sect. 1.3.3.1.
- SW VERSION—The version designation consists of the main version (byte 4), minor version (byte 5), and patch designation (byte 6). The last byte can be used as an additional sign.
- DATE—production date of the measuring module. It consists of the number of the month (byte 14) and the number of the year (bytes 14 and 15).

- MultiSlave ID—unique identifier of the MultiSlave module. Address: 16, length: 8B
- CIPH—flag whether the module supports encrypted communication. Address: 26, length: 1B
- 1W—software limitation for the maximum number of 1-wire bus sensors, if used. Address: 27, length 1B
- FAMILY—type designation of the measuring module/MultiSlave. The type markings are the names of the final solutions as described in Chap. 2. Address: 28, length: 4B
- MultiSlave name—the name of the measuring device, as it will be indicated in the software applications that communicate with the measuring module. Address: 32, length: 8B
- FW VERSION—designation of the uBUS protocol implementation version. The labeling is the same as in the SW VERSION section.

The uBUS application protocol uses a number format with a fixed decimal point for numeric values. This format is referred to as $Q(m, n)$, where $m+n$ is the number of bits used to represent the value notation, n defines the resolution (2^{-n}), and m defines the range $(-2^m \ldots 2^m - 2^{-n})$. A constant length of 32 bits is used when representing a real number, while the Q parameter indicates the used range and precision.

In addition to the configuration of the multislave module, the configuration for all the sub-slaves must be implemented. Figure 1.21 shows the memory map for one sub-slave. The configuration of the first sub-slave starts at address 64. The configuration of the i-th sub-slave ($i = 0 \ldots 15$) is at address $64 + 3 \cdot i$. The length of this configuration is 24B. The meaning of individual parts is as follows:

- TYPE—specification of the sub-slave device type. The format and individual parts of the type specification are described in the uBUS protocol specification section (Chap. 1.3.3). Length: 3B.
- Q—numeric format specification for MIN VALUE/MAX VALUE values. The specification is given in Table 1.7.
- UNIT—a physical unit of the measured quantity. The specification for the physical unit is described in Sect. 1.3.3.5. Length: 4B.
- SUBSLAVE NAME—the name of the module as it will be displayed in the supporting software. Length: 8B.
- MIN VALUE/MAX VALUE—information about the minimum and maximum values the connected sensor can measure. Values are stored in the Q numeric format.
- COEF 1 (2,3)—coefficients for converting the raw measured value to the resulting value. Each coefficient has its Q format defined. The use of coefficients is optional. If coefficients are used, the value conversion takes place in the service software, where the formula for converting this value is also defined.

64+ 3*i+	0	1	2	3	4	5	6	7	8	9	10	11	12	14	14	15
0	TYPE			Q	UNIT				-	-	-	-	-	-	-	-
1	SUBLASVE NAME							MIN VALUE				MAX VALUE				
2	Q1	COEF 1			Q2	COEF 2			Q3		COEF 3					-

Fig. 1.21 Sub-slave memory map configuration

Table 1.7 Format flag for fixed decimal point

Format	Notation	Accuracy	Range
0	$Q(31,0)$	2^{-31}	$\langle-1, 1\rangle$
1	$Q(24,7)$	2^{-24}	$\langle-128, 128\rangle$
2	$Q(8,23)$	2^{-8}	$\langle-9.38e6, 9.38e6\rangle$
3	$Q(0,31)$	2^{-0}	$\langle-4.291e9, 4.291e9\rangle$

1.3.3 Specification of the uBUS Protocol

The uBUS application protocol was designed to exchange sensory data. Therefore, when specifying it, emphasis was placed on information related to the measured values themselves. The primary entity to which all information is linked is the sensor. Additional information is associated with this entity:

- Sensor type specification, which consists of the following:

  - Specifications of the physical unit in which the sensor produces data (*TypeGroup1*)
  - Specification of the principle of acquiring the measured value—the physical principle of measurement (*TypeGroup2*)
  - Serial number of the sensor

In further processing, at the level of the sensory information system, the following information is assigned to the sensor:

- Exact measurement time
- Place of measurement
- Sensor location
- Authorizations for displaying values.

TypeGroup1 and *TypeGroup2* are codebooks of fundamental physical quantities and measurement methods. Table 1.8 contains a list of the main types defined by the uBUS protocol. These types are listed together with the basic physical unit.

In Table 1.9, there is a list of physical principles or methods of measurement used by individual sensors.

The third part of the sensor type specification is the sensor serial number for the given physical quantity and measurement method. Table 1.10 shows a specific type specification for a selection of available sensors.

Table 1.8 Sensor type specification—TypeGroup1

ID	Value	Unit	ID	Value	Unit
1	Length	m	53	Magnetic induction	T
2	Mass	kg	54	Electric induction	C/m^2
3	Time	s	81	Momentum	kg*m/s
4	Electrical current	A	82	Moment of inertia	$kg*m^2$
5	Temperature	°C	83	Moment of momentum	$kg*m^2/s$
6	Lath quantity	mol	84	Volumetric speed	m^3/s
7	Luminosity	cd	85	Speed	m/s
16	Speed	m/s	86	Acceleration	m/m^2
17	Plane angle	rad	87	Force, weight	N
18	Spatial angle	sr	88	Moment of force	N*m
19	Boolean value		89	Surface tension	N/m
20	Counter	n	90	Pulse of force	N*s
21	Precent	%	97	Density	kg/m^3
22	Direction		98	Volume	m^2
33	Electric current density	A/m2	99	Areal content	m^2
34	Electric charge	C	100	Radiation dose	Gy
35	Irradiation	C/kg	101	Radiation dose rate	Gy/s
36	Capacity	F	102	Pressure	Pa
37	Frequency	Hz	113	Heat capacity	J/K
38	Radiation energy, electrical work	j	114	Specific heat	J/(kg*K)
39	Electrical resistance	Ω	129	Bright	cd/m^2
40	Electrical conductivity	S	130	Luminous flux	lm
41	Electrical voltage	V	131	Luminous flux density	lux
42	Power	W	132	UV index	–
43	Radiation intensity	W/m^2	145	Angular velocity	rad/s
49	Magnetic field strength	A/m	146	Angular acceleration	rad/s^2
50	Permittivity	F/m	161	Sound pressure level	dB
51	Inductance	H	162	Relative humidity	%
52	Permeability	H/m	163	Strain gauge extension	uStrain
53	Magnetic induction	T	177	Memory capacity	B
54	Electric induction	C/mm^2	193	From datalogger	–
81	Momentum	kg*m/s	209	Communication bus bridge	–
225	System device	–			

The uBUS protocol defines the primary division of communication into unicast and broadcast communication. According to this division, individual functions of the protocol are also divided into:

1. Unicast function

   a. Diagnostic
   b. Configuration
   c. For data access

Table 1.9 Sensor type specification—TypeGroup2

ID	Text	ID	Text
1	Not defined	37	Capacitive
17	Contact	38	Piezoelectric
18	Spring-loaded	49	IR light measurement
19	Movement of mechanical parts	50	Visible light measurement
20	Bimetallic	51	UV light measurement
21	Rotary	65	Ultrasonic waves measurements
22	Sliding	81	Elemental element detector
23	Weight measurement	97	I/O device
24	Pressure	98	Pulse counter
33	Electrical resistance measurement	99	ROM memory
34	Electrical current measurement	100	RWM memory
35	Electrical voltage measurement	113	Direct to digital temperature
36	Induction	114	Direct to digital humidity

Table 1.10 Sensor type specification—TypeGroup3

ID1	ID2	ID3	Text
2	33	1	Tensometric module TNZ
5	33	2	Platinum thermometer PT100
5	33	3	Platinum thermometer PT500
5	33	4	Platinum thermometer PT100
5	113	6	Digital thermometer DS18S20
5	113	7	Digital thermometer DS18B20
5	113	8	Digital thermometer DHT22
5	113	9	Digital thermometer MCP 9808
5	113	10	Digital thermometer LM 75
131	49	1	Sensor SI 1145—IR light
131	50	1	Sensor SI 1145—visible light
132	51	1	Sensor SI 1145—UV light

2. Broadcast function

   a. Configuration
   b. Access to measurement functions.

Table 1.11 contains a summary of the supported commands, their symbolic designation, and numeric code.

Unpredictable errors may occur during communication which the measurement module must report. Standard MODBUS protocol codes (Table 1.3) are used to report errors. In addition to them, additional error codes are also defined—Table 1.12.

In the following text, we present a detailed description of the functions of the uBUS protocol. The meaning of abbreviations used in this section is:

Table 1.11 uBUS unicast functions

Type	Dest	Function code	Description	Function name
DIAGNOSTIC	Device	0x01	List of active slave devices	CMD_DEV_SLAVES
		0x02	Echo response	CMD_DEV_ECHO
		0x03	Device unique identifier	CMD_DEV_ID
		0x04	Information on device	CMD_DEV_INFO
	Slave	0x05	Device type	CMD_SLAVE_INFO
		0x06	Sensor's measurement unit	CMD_SLAVE_UNIT
		0x07	Returns the current sensor state	CMD_SLAVE_STATE
		0x08	Returns information on sensor	CMD_SLAVE_INFO
CONFIGGURATION	Slave	0x0A	Reset	CMD_SLAVE_RESET
		0x0B	Set parameter	CMD_SLAVE_SET_PARAM
		0x0C	Load the parameter	CMD_SLAVE_GET_PARAM
	Device	0x10	Start—initiates the action (measurement)	CMD_DEV_START
		0x11	Stop—stops the action (measurement)	CMD_DEV_STOP
		0x12	Reset	CMD_DEV_RESET
		0x13	Parameter/value setting	CMD_DEV_SET
		0x14	Parameter/value loading	CMD_DEV_GET
		0xE1	Initialization of encrypted communication	CMD_SECURE_INIT
		0xE2	Communication Request to send public key	CMD_SECURE_ENCODE_KEY
DATA	Slave	0x15	Set the value	CMD_SLAVE_GET_VALUE
		0x17	Load the value	CMD_SLAVE_SET_VALUE

Table 1.12 uBUS error codes

Exception code	Title	Meaning
0x10	ERR_PARAMETER_NOT_ IMPLEMENTED	Settingloading the parameter from/to the slave device—parameter not implemented
0x11	ERR_PARAMETER_NOT_SET	Setting the parameter to the slave device—the parameter could not be set
0x12	ERR_PARAMETER_VALUE_ERROR	Parameter setting of the slave device—parameter value not allowed
0x21	ERR_IO_INPUT	Error loading data—the device is output
0x22	ERR_NO_VALUE	Error loading data—no value ready
0x25	ERR_FLAG_DATA_FORMAT	Error in request parameter—incorrect sub-function additional data
0x29	ERR_FLAG_NOT_SUPPERTED	Error in request parameter—the device does not support the requested sub-function of the command
0x30	ERR_INTERFACE_VALUE	The addressed sensor is an interface (1-wire) and does not support the FLAG_DATA_SINGLE flag. Use a different method to retrieve values from the slave sensors
0xE0	ERR_DATA_FORMAT	An encryption layer is required for communication

- FC—byte at position 2 in the request and response of the uBUS communication frame
- EC—Error Code. Error code according to Tables 1.3 and 1.12

Calculation CRC Recall that each uBUS device has a 2-byte address, which is, in the following text, referred to as 0xNNN0. The specific sensor address consists of the measuring module address and the sensor within the measuring module address (Fig. 1.8). The address of the hardware module, therefore, always ends with the value 0.

The CRC16 algorithm—Listing 1.1—with the polynomial value 0xA001 is used as the verification algorithm. The algorithm given in [5] as a fast algorithm with complexity $O(n)$ (Listing 1.1) can be used for the calculation of the checksum itself. At the same time, it is necessary to have the pre-calculated values for the look-up table of size 512 B.

```
void calc_crc(uint8_t *buffer,uint8_t length) {
  uint8_t value1 = 0xFF;
  uint8_t value2 = 0xFF;
  uint16_t index = 0x00;
  // Look-up table: crcHi, crcLo
  for(uint8_t i=0 ; i<length ; i++){
    index  = value2 ^ buffer[i];
    value2 = value1 ^ crcHi[index];
    value1 = crcLo[index];
  }
  uBus.CRC16_L = value1;
  uBus.CRC16_H = value2;
}
```

Listing 1.1 Algorithm for fast calculation of CRC16

Table 1.13 Low-level data format for uBus frame

Index in the data frame	0	1	2	3
Data format for 0x12345678	0x78	0x56	0x34	0x12

Table 1.14 The format of the communication packet of the uBUS protocol

Byte	0	1	2	$3..(n-3)$	$n-2$	$n-1$
Meaning	Address MSB	Address LSB	Function code	Data	CRC MSB	CRC LSB

Error Codes In the case of an error response, the value 0x80 (error flag) is always added to the response function code. This way of working with the function code allows easy differentiation: valid function codes are from 0x01 to 0x7F. Other values represent "function code + error flag." During the communication, it is possible to distinguish the status of the processed function:

- In the case of error-free processing of the requested function, the response of the slave node contains a copy of the executed function code.
- In the event of an error during function processing, the slave node's response contains the value FC+0x80 as the function code.

Data Format The request format of individual functions is given in the following content. The Little Endian format is used for data longer than 1B. An example of writing the value 0x12345678 in the Little Endian format is given in Table 1.13.

In the following text, we present the basic specification of the uBus application protocol. The uBUS application protocol defines the basic format of the communication packet, which consists of the destination address (2B), FC—Function code (1B), the data part (mandatory part), and the checksum (last 2 bytes). This

Table 1.15 uBUS response format

Action/byte	0,1	2	3	4
Request	0xNNNN	FC	(DATA)	
Response OK	0xNNNN	FC	DATA	
Response ERROR	0xNNNN	FC+0x80	EC	

frame format is indicated in Table 1.14. The value *n* represents the length of the communication frame. Data is sent in binary unless otherwise stated.

In the following text, we will work with this format. For clarity, the last 2 bytes (CRC) will be omitted in the following tables. The meaning of the address of the target module is shown in Fig. 1.8. In the following text, the marking for the target address of the measuring module is used:

- *0xNNN0*—it represents the address of the MultiSlave module.
- *0xNNNN*—it represents the address of any MultiSlave submodule.
- *0xNNNn*—*n* represents a submodule with an internal address of 0xn in a MultiSlave module with an address of *0xNNN0*.
- *0xNNNx*—the value of *x* does not matter. The message with this address is always processed as a command for the MultiSlave. It is not a specific submodule.

The principle of communication in the uBUS protocol is the same as in the MODBUS protocol. Table 1.15 shows an example of a request, a positive, and an error response. The communication rules can be summarized in a few points:

- The address of the destination module has 2B
- FC—the code of the requested function to be performed by the addressed module
- DATA—optional part
- The response contains a copy of the original address
- The FC value in the response has a value:

  - A copy of the FC in case of a positive response
  - In case of an error response, the value 0x80 is added to the response: (FC+0x80).

- In the error response, the value EC (Error Code) is at the third position in response.

1.3.3.1 Command CMD_DEV_SLAVES (0x01)

Command returns the bit mask of the active devices of the MultiSlave device (Table 1.16). A MultiSlave module can contain a maximum of 16 submodules . The submodule with the address 0x0F is reserved for implementing the internal EEPROM memory, where the device configuration is stored. The memory map of

Table 1.16 Specification of the data frame for the command CMD_DEV_SLAVES

Action/byte	0	1	2	3	4
Request	0xNN	0xN0	0x01		
Response OK	0xNN	0xN0	0x01	DATA	

Table 1.17 The meaning of the DATA part in the answer CMD_DEV_SLAVES

	Byte 3								Byte 4							
Bit	7	6	5	4	3	2	1	0	7	6	5	4	3	2	1	0
Sub-slave device	S_{15}	S_{14}	S_{13}	S_{12}	S_{11}	S_{10}	S_9	S_8	S_7	S_6	S_5	S_4	S_3	S_2	S_1	S_0

Table 1.18 Specification of the data frame for the command CMD_DEV_ECHO

Action/byte	0	1	2	3	..	$n-3$
Request	0xNN	0xNx	0x02	DATA		
Response OK	0xNN	0xN0	0x02	DATA		

the MultiSlave node is in Fig. 1.19 and Fig. 1.20. The destination address is always the MultiSlave address (0xNNN0).

The meaning of individual bits in response is in Table 1.17, where S_i has the value:

- 1—if there is a slave at address i on a given device LBS
- 0—if no device at the address I is connected

1.3.3.2   Command CMD_DEV_ECHO (0x02)

Echo command. The response will contain an exact copy of the data as in the request. The CRC checksum is not copied but calculated, Table 1.18.

- Address *0xNNNx*—the command is addressed to the device, not the sensor. The slave device must process the request for arbitrary n.
- The answer will always contain the address *0xNNN0*, i.e., the device's address.
- The data part can be 1 to 253 B long. The content of the data part is arbitrary. The response will contain a copy of the data part.
- The data part can be 1 to 253 B long. The content of the data part is arbitrary. The response will contain a copy of the data part.
- The CRC checksum is recalculated when the response is sent.

Table 1.19 Specification of the data frame for the command CMD_DEV_ID

Action/byte	0	1	2	3	..	10
Request	0xNN	0xNx	0x03			
Response OK	0xNN	0xN0	0x03	DATA		

Table 1.20 Specification of the data frame for the command CMD_DEV_INFO

Action/byte	0	1	2	3	4	5	6	7
Request	0xNN	0xNx	0x04	INFO_PART				
Response OK	0xNN	0xN0	0x04	INFO_PART	DATA			

1.3.3.3   Command CMD_DEV_ID (0x03)

The command returns the unique 8B identifier of the measuring device. It is the command for the MultiSlave module. The response will always contain the address *0xNNN0*, i.e., the address of the device, Table 1.19.

1.3.3.4   Command CMD_DEV_INFO (0x04)

Command returns specific information about the status of the measurement module. The INFO_PART flag gives the choice of information type. The length of the data part in the response is 4 B unless otherwise specified (Table 1.20).

The response contains data part 4B—in positions 4 to 7. In the following text, the communication packet is marked as data_frame.

Meaning INFO_PART

FLAG_DEV_INFO_SW   (0x01) The version number of the application firmware. This designation is linked to the version of a specific application, not the version of the uBUS protocol implementation itself. The version tag consists of a major version (data[4]), a minor version (data[5]), and a patch tag (data[6]). The last bit can be used as an additional sign.

FLAG_DEV_INFO_HW   (0x02) The version of the hardware solution of the measuring module. No version numbering rules are defined.

FLAG_DEV_INFO_RELEASE   (0x03) Date of issue/installation of the measuring module:

- data_frame[4]—0x0
- data_frame[5]—serial number of the month: 0x0 (January)—0xB (December).
- data_frame[6...7]—year. The year 2022 is given in hexadecimal format (0x07E6), resulting in data_frame[6] = 0x07, data_frame[7] = 0xE6.

FLAG_DEV_INFO_RSA   (0x04) Information on communication encryption. The response is 1 byte long.

- data_frame[7] = 0x1—communication must be encrypted.
- data_frame[7] = 0x0—communication is not encrypted.

FLAG_DEV_INFO_NAME   (0x05)

- The answer has 8 bytes. The response format is ASCII characters.

FLAG_DEV_INFO_FAMILY   (0x06) The designation of the family or type of the measuring module. The modules labeled OWB, TNZ, RRM, and others are described in the following section. The data format is ASCII characters.

- data_frame[4...6]—3-character type designation of the device
- data_frame[7]—release version (optional)

FLAG_DEV_INFO_1W_SENSORS   (0x08) The number of existing sensors on the 1-wire bus if the device has a 1-wire bus.

- The response is 1 byte long.
- data_frame[4]—the numerical value of the answer.

FLAG_DEV_INFO_FW   (0x09) Information on the firmware version of the measuring module. No version numbering rules are defined.

FLAG_DEV_INFO_STATE   (0x0A) Information about the status of the measuring board. The answer is 1 byte. Possible values are:

- DEVICE_STATE_STOP (0)—measurements on the module are stopped
- DEVICE_STATE_RUN (1)—measurements on the module are running, and the module is active
- DEVICE_STATE_LOW_VOLTAGE_STOP (2)—low supply voltage, measurements are stopped.
- DEVICE_STATE_LOW_VOLTAGE_RUN (3)—low supply voltage, measurements are suspended.

FLAG_DEV_INFO_1W_SENSORS_MAX   (0x0B) The maximum number of sensors allowed on the 1-wire bus. This value is written in the configuration of the measuring module. The answer is 1 B. Possible values are 1—255.

1.3.3.5   Command CMD_SLAVE_UNIT (0x05)

Command returns 3 bytes that specify the sensor type, Tables 1.21. The sensor type is specified according to Tables 1.8, 1.9, and 1.10.

Table 1.21 Specification of the data frame for the command CMD_SLAVE_TYPE

Action/byte	0	1	2	3	4	5
Request	0xNN	0xNN	0x05			
Response OK	0xNN	0xNN	0x05	TYP_1	TYP_2	TYP_3

1.3.3.6   Command CMD_SLAVE_UNIT (0x06)

Command returns 4B which it defines the measured physical unit according to the SI system (Table 1.22).

Each response byte in the data part consists of two-byte quads that define one basic physical unit. Coding four bits for each physical unit is a 4-bit complementary code that talks about its power in the resulting unit. The units are given in the order:

- DATA[3] ⇒ m
- DATA[4] ⇒ kg and s
- DATA[5] ⇒ A and K
- DATA[6] ⇒ mol and cd (for details, see Table 1.23)

where

- s—length, unit: meter [m]
- m—mass, unit: kilogram [kg]
- t—time, unit: second [s]
- I—electric current, unit: Ampere [A]
- T—temperature, unit: degrees Kelvin [K]
- n—the amount of substance, unit: mol
- J—luminosity, unit: candela [cd]
- X—not used

1.3.3.7   Command CMD_SLAVE_STATE (0x07)

Command returns 1B with status information about the measuring module, Table 1.24.

Each measurement module must inform about its status or the readiness of the measured data. Table 1.25 contains a list of available states. The use of this command before the request for measured data (Table 1.36) is not mandatory. The meaning of the data part of the answer is given in Table 1.25.

Table 1.22 Specification of the data frame for the command CMD_DEV_ID

Action/byte	0	1	2	3	4	5	6
Request	0xNN	0xNx	0x06				
Response OK	0xNN	0xN0	0x06	DATA[3]	DATA[2]	DATA[1]	DATA[0]

Table 1.23 The meaning of the data part of the command CMD_SLAVE_UNIT

	Byte 3								Byte 4							
Bit	7	6	5	4	3	2	1	0	7	6	5	4	3	2	1	0
Meaning	X	X	X	X	s_3	s_2	s_1	s_0	m_3	m_2	m_1	m_0	t_3	t_2	t_1	t_0
	Byte 5								Byte 6							
Bit	7	6	5	4	3	2	1	0	7	6	5	4	3	2	1	0
Meaning	I_3	I_2	I_1	I_0	T_3	T_2	T_1	T_0	n_3	n_2	n_1	n_0	J_3	J_2	J_1	J_0

Table 1.24 Specification of the data frame for the command CMD_SLAVE_STATE

Action/byte	0	1	2	3
Request	0xNN	0xNN	0x07	
Response OK	0xNN	0xNN	0x07	DATA

Table 1.25 Meaning of the data part of the answer

Value	Symbolic name	Meaning
0x00	STATE_READY	The measured value ready for reading
0x01	STATE_NO_DATA	The measured value is not available
0x02	STATE_SENSOR_BUSY	Sensor busy—measurement has started
0x03	STATE_DATA_PROCESSING	Sensor busy—processes the measured data

Table 1.26 Specification of the data frame for the command CMD_SLAVE_INFO

Action/byte	0	1	2	3	4	..	..	11
Request	0xNN	0xNx	0x08	INFO_PART				
Response OK	0xNN	0xN0	0x08	INFO_PART	DATA (8B)			

1.3.3.8 Command CMD_SLAVE_INFO (0x08)

The command returns the specific properties of the addressed submodule (Table 1.26). Supported parts are submodule name and maximum and minimum value of the measurement range for the submodule sensor (Table 1.27).

1.3.3.9 Command CMD_SLAVE_RESET (0x0A)

The command resets the sensor settings. This is a software reset or a reinitialization of the submodule. The measurement submodule must initialize the addressed sensor in the same way as when starting the entire module. The format of the request is given in Table 1.28.

Note that the response contains the SC code instead of the data part:

- 0x80: State OK
- 0x81: Long time operation

Table 1.27 Specification of the data frame for the command CMD_SLAVE_INFO

INFO_PART	Symbolic name	Meaning
0x01	FLAG_SENSOR_NAME	Sensor/module name. Sensor names in different measuring devices may be repeated. Thus this is not a unique name
0x02	FLAG_SENSOR_MIN_VALUE	The lower range of the sensor. The value is given as an integer in the format Qm.n. The format is as follows: "0 0 0 F M M M M," where F is the data format (Qm.n) and M is the value itself as a 4B number. The format is defined as • F=0xFF, F=0x0: format = Q16.15 • F=1, format ⇒ Q0.31 • F=2, format ⇒ Q8.23 • F=3, format ⇒ Q16.15 • F=4, format ⇒ Q24.7 • F=5, format ⇒ Q31.0
0x03	FLAG_SENSOR_MAX_VALUE	The upper range of the sensor. The format of the value is the same as in the case of the part INFO_SENSOR_MIN_VALUE

Table 1.28 Specification of the data frame for the command CMD_SLAVE_RESET

Action/byte	0	1	2	3
Request	0xNN	0xNN	0x0A	
Response OK	0xNN	0xNN	0x0A	SC

Table 1.29 Specification of the data frame for the command CMD_SET_PARAM

Action/byte	0	1	2	3	4	5
Request	0xNN	0xNN	0x0B	ParamName	ParamValue	
Response OK	0xNN	0xNN	0x0B	SC		

1.3.3.10   Command CMD_SLAVE_SET_PARAM (0x0B)

The command sets the parameter value of the sensor or measuring module. These are specific parameters of selected measurement modules. The command format is given in Table 1.29.
SC—State Code in response:

- 0x80—State OK
- 0x81—Long time operation

Specific error conditions. EC values in the error response:

- 0x03—ILLEGAL_DATA_VALUE. Error in the request. No parameter was specified.
- 0x10—PARAM_NOT_IMPLEMENTED. The parameter not implemented

- 0x11—PARAM_NOT_SETTED. The parameter could not be set.
- 0x12—ERR_PARAMETER_VALUE_ERROR. Illegal parameter value

Slave module parameters are optional. Universal parameters are:

- 0x06—PARAM_MEASURE_REPEAT. The number of repeated readings of values from the sensor during the measurement. The default value is 1.

For the OWB (One-Wire Booster) module, some parameters set the timing parameters of the 1-wire bus. The meaning of these timing parameters is described in Sect. 2.1. The parameters of the OWB module refer to Table 1.29, where the meaning of the individual parameters is given. Unless otherwise stated, values are given in microseconds. The meaning of the following parameters is described in detail in Fig. 1.26 and Table 1.41.

- 0x14—PARAM_1W_CONSTANT_RST_MASTER. Sets the parameter RESET_PULSE—reset pulse length.
- 0x15—PARAM_1W_CONSTANT_RST_PRESENCE. Sets the parameter READ_PRESENCE—bus status reading time for detecting the presence of a slave on the 1-wire bus.
- 0x16—PARAM_1W_CONSTANT_RST_DELAY. Sets the parameter RESET_DELAY—time until the end of the slot for the RESET operation.
- 0x17—PARAM_1W_CONSTANT_W_ZERO. Sets the parameter WRITE_ZERO—time to write bit 0 to 1-wire bus.
- 0x18—PARAM_1W_CONSTANT_W_ONE. Sets the parameter WRITE_ONE—time to write bit 1 to 1-wire bus.
- 0x19—PARAM_1W_CONSTANT_W_ONE_DELAY. Sets the parameter WRITE_ONE_DELAY—time until the end of the communication slot after writing bit 1.
- 0x1A—PARAM_1W_CONSTANT_W_DELAY. Sets the parameter WRITE_DELAY—time until the end of the communication slot after writing one bit.
- 0x1B—PARAM_1W_CONSTANT_R_PULSE. Sets the parameter READ_PULSE—time for setting the reading pulse length.
- 0x1C—PARAM_1W_CONSTANT_R_PRESENCE. Sets the parameter READ_PRESENCE—time for setting the moment of reading the value from the 1-wire bus. This parameter is critical. If the setting is incorrect—too short or too long—the 1-wire bus master device can evaluate the bus status incorrectly.
- 0x1D—PARAM_1W_CONSTANT_R_DELAY. Sets the parameter READ_DELAY—time until the end of the communication slot after adding one bit. The value of OW_BOOSTER further reduces this time.
- 0x1E—PARAM_1W_CONSTANT_OW_BOOSTER. Sets the parameter OW_BOOSTER—time for activation of the BOOST mode after communication. At this time, a hard 5 V is supplied to the 1-wire bus for a quick return of the voltage level to the idle state, i.e., log.1 level.
- 0x1F—PARAM_1W_CONSTANT_TIME_SLOT. The additional parameter of 1-wire bus. The value is given in microseconds. The parameter can only be read. It is not used for timing, but in the parent application, the time for transferring 1 bit is determined according to it.

Table 1.30 Specification of the data frame for the command CMD_SLAVE_GET_PARAM

Action/byte	0	1	2	3	4	5
Request	0xNN	0xNN	0x0C	ParamName		
Response OK	0xNN	0xNN	0x0C	ParamName	ParamValue	

Table 1.31 Specification of the data frame for the command CMD_DEV_START

Action/byte	0	1	2	3
Request	0xNN	0xNN	0x10	
Response OK	0xNN	0xNN	0x10	SC

- 0x20—PARAM_1W_DS_CONVERT_DELAY. The waiting time required for the conversion of the measured value. It is given as an integer in ms. Allowed range: 256–1000 ms.
- 0x21—PARAM_1W_TIMMING_SETTINGS. 1-wire bus timing setting. All the 1-wire bus parameters (0x14 - 0x1E) are set according to the predefined constants stored in the configuration memory of the measuring module. Allowed values are:

  1. Standard timing
  2. Timing for a long bus
  3. user-set timing

1.3.3.11   Command CMD_SLAVE_GET_PARAM (0x0C)

The command reads the parameter value of the sensor or measuring module. The parameter list is the same as for the CMD_SLAVE_SET_PARAM command. The request and response formats are given in Table 1.30.

Specific error conditions:

- 0x03—ILLEGAL_DATA_VALUE. Wrong value of the set parameter
- 0x10—PARAM_NOT_IMPLEMENTED. Parameter not implemented

1.3.3.12   Command CMD_DEV_START (0x10)

The command enables the periodic measurement on the measuring module. This command applies to all the existing submodules. The measurement frequency must be set using the setParam command, with the DEV_PARAM_SAMPLE_TIME parameter (if the measurement module supports it). The request and response formats are given in Table 1.31.

Table 1.32 Specification of the data frame for the command CMD_DEV_STOP

Action/byte	0	1	2	3
Request	0xNN	0xNN	0x11	
Response OK	0xNN	0xNN	0x11	SC

Note:

- The response contains the SC code instead of the data part:
  - 0x80—State OK
  - 0x81—Long time operation
- In the error response, the EC status code specifies the errors:
  - 0x01—Illegal function, unsupported feature.

1.3.3.13 Command CMD_DEV_STOP (0x11)

The command stops the measurement started by the CMD_DEV_START command. The request and response formats are given in Table 1.32.
Note:

- The response contains the SC code instead of the data part:
  - 0x80—State OK
  - 0x81—Long time operation
- In the error response, the EC status code specifies the errors:
  - 0x01—Illegal function, unsupported feature.

1.3.3.14 Command CMD_DEV_RESET (0x12)

The command starts a reset of the measuring module. The reset can be software and hardware. The software reset (Table 1.33, request 1) performs the following actions:

- Starts reset of all the sensors
- Initializes the values read off of the configuration memory in the same way as after the HW reset
- Sets the measurement module to a state when the measurements are allowed

The hardware reset (Table 1.33, a request 2)—the request differs in that the data part contains the module's to be reset address. It is a copy of the first 2 bytes of the request. During the hardware reset, the module is physically reset.

Table 1.33 Specification of the data frame for the command CMD_DEV_RESET

Action/byte	0	1	2	3	4
Request	0xNN	0xNN	0x12		
Response OK	0xNN	0xNN	0x12	ADDR[0]	ADDR[1]

Table 1.34 Specification of the data frame for the command CMD_SLAVE_GET_PARAM

Action/byte	0	1	2	3	4..	..7
Request	0xNN	0xNN	0x13	ParamName	ParamValue	
Response OK	0xNN	0xNN	0x13	SC		

1.3.3.15   Command CMD_DEV_SET (0x13)

It is the command to set the measuring module's global parameter/value Table 1.34. This is a universal setting command. The parameters, which can be set, may not be supported by all the measurement modules.
Parameters of the function CMD_DEV_SET:

- 0x02—DEV_PARAM_SAMPLE_TIME. In the case of continuous measurement, it defines the interval for automatically starting the measurement. It is given in ms. This value applies to all the existing submodules in the MultiSlave module.
- 0x03—DEV_PARAM_TELEMETRY. Read-only data—gauge board telemetry. The data part of the response is 4B. Each byte represents one telemetry value. If the measurement module does not measure a specific quantity, 0x0 will be indicated in the response. The meaning of individual bytes in the response is:

  - data[4]: voltage U1 [V]—supply voltage of the measuring module
  - data[5]: voltage U2 [V]—supply voltage of a specific peripheral
  - data[6]: current I1 [mA]—current consumption of the measuring module
  - data[7]: current I2 [mA]—current consumption of a specific peripheral. It depends on the specific measuring module. For the OWB module, the current consumption of the 1-wire bus is given.
  - The meaning of the values U1, U2, I1, and I2 is not strictly defined. However, it is recommended that U1 and I1 relate to the supply voltage and current electricity consumption, respectively. The voltage U2 and current I2 are recommended for implementation for the external bus monitoring if, for example, the 1-wire bus is used.

Specific error states:

- 0x10—PARAM_NOT_IMPLEMENTED—parameter not implemented
- 0x03—ILLEGAL_DATA_VALUE—parameter value is out of range
- 0x12—ERR_PARAMETER_VALUE_ERROR—failed to load the required parameter

Table 1.35 Specification of the data frame for the command CMD_DEV_GET

Action/byte	0	1	2	3	4	5
Request	0xNN	0xNN	0x14	ParamName		
Response OK	0xNN	0xNN	0x14	ParamName	ParamValue	

Table 1.36 Specification of the data frame for the command CMD_SLAVE_SET_VALUE

Action/byte	0	1	2	3	4..	..21
Request	0xNN	0xNN	0x15	FLAG	DATA	
Response OK	0xNN	0xNN	0x15	SC		

1.3.3.16 Command CMD_DEV_GET (0x14)

The command serves for loading the global parameter/value for the measurement module. The parameter list is the same as for the CMD_DEV_SET function. The format of the communication packet is given in Table 1.35.
Specific error conditions:

- 0x10—PARAM_NOT_IMPLEMENTED. Parameter is not implemented.

1.3.3.17 Command CMD_SLAVE_SET_VALUE (0x15)

The command serves for writing the value into the measuring module. The function is supported only for measurement modules that contain output submodules. This function is also used for saving the configuration of the measuring board in the EEPROM memory. This document does not define the configuration memory implementation. The configuration memory size is 1 kB. The format of the communication packet is given in Table 1.36.

- The response contains the SC status code in the data part:
  - 0x80—State OK
  - 0x81—Long time operation. The operation will be processed for longer than the time required to send the response.
- In the error response, the EC status code specifies the errors:
  - 0x21—ERR_IO_INPUT. Error when loading data—this is the output device
  - 0x27—ERR_DATA_WRITE. Registration failed
  - 0x28—ERR_DATA_FORMAT. Data is in the wrong format

The data part size is 2B unless otherwise specified. It can vary between 1 B and 16 B, according to the set format. Values of supported functions—FLAG attribute:

- 0xAE—FLAG_DATA_MEMORY_CLEAR. Deleting the device configuration. After deleting the configuration, the predefined values for all the sub-slave modules are used.

  - In the data part of the request (data[4–5]), the address of the module (data[1–2]) is indicated to prevent unwanted deletion of the module configuration.

- 0xAF—FLAG_DATA_MEMORY. Writing user data to FLASH. Since it is a FLASH memory, data can be written in pages, which, in most cases, have a size of 1kB. Only one writing to the memory (to a certain page) is possible after the deletion by the command FLAG_DATA_MEMORY_CLEAR. It is not possible to overwrite the memory. If it is necessary to modify the data, which has already been written, it is necessary to wipe out the entire configuration memory. The configuration memory is organized in rows of 16B, and the number of rows is 64 (Figs. 1.19 and 1.20). Therefore, the data is sent in the 16 B size. The data part contains the request:

  - Address or the line number in data[4] configuration memory. Values 0 . . . 63 are allowed.
  - The data itself: data[5–21]. Data of the 16-byte size is to be written to the requested address in the configuration memory

There are specific functions for the OWB hardware module. This module contains a driver for the 1-wire bus, to which up to 128 sensors can be connected. The module must be able to communicate with each sensor. The following variants of the CMD_SLAVE_SET_VALUE function are available for this communication. The data part of each function has a variable length. The FLAG part (data[3]) refers to the CMD_SLAVE_SET_VALUE command.

FLAG = 0x20 (FLAG_DATA_1W_SENSOR_ADDRESS)
It sets the sensor ID specified in the data part as the active sensor to be communicated with on the 1-wire bus. The data[4–12] in the request represents the 8-byte address of the requested 1-wire sensor. This code can also be used in the CMD_SLAVE_GET_VALUE function to find the currently selected 1-wire sensor. Error responses:

- ILLEGAL_DEVICE_ADDRESS (0x02)—the sensor with the given address does not exist on the 1-wire bus.

FLAG = 0x21 (FLAG_DATA_1W_SENSOR_INDEX)
It sets the active 1-wire sensor for communication. There is a list of allowed sensors in the configuration memory. This command sets the ID of the i-th 1-wire sensor, whose identifier (8B ID) is stored in the FLASH memory as the i-th sensor in the sequence. An internal list of 1-wire sensors is generated when the OWB device is turned on and at each device reset request. In the request, data[5] represents the sequence number of the sensor as detected on the 1-wire bus. The sensor identifiers determine sensor sorting, which any setting cannot influence. The first detected sensor is recorded with an index of 0. This code can also be used in the

CMD_SLAVE_GET_VALUE function to find the index of the currently selected 1-wire sensor. Error responses:

- ILLEGAL_DEVICE_ADDRESS (0x02)—the sensor with the given index does not exist on the 1-wire bus.

FLAG = 0x28 (FLAG_DATA_1W_CONVERT_BULK)

Starting temperature conversion on 1-wire bus sensors. The conversion lasts from 150 ms to 750 ms, depending on the precision setting of the individual sensors. This function is non-blocking. The OWB module sends a response with a LONG_TIME_OPERATION—0x81 response. During the conversion, the OWB module must respond to any command. The measured value will be ready only after the time required for the value conversion has elapsed.

FLAG = 0x29 (FLAG_DATA_1W_CONVERT_SINGLE)

It starts the temperature conversion in SINGLE mode on the sensor that is set as active.

The given sensor can be changed with the FLAG_DATA_1W_SENSOR_ADDRESS and FLAG_DATA_1W_SENSOR_INDEX functions. The OWB module responds the same as for the FLAG_DATA_1W_CONVERT_BULK function.
Error responses:

- ILLEGAL_DEVICE_ADDRESS (0x02)—if there is no sensor on the bus with an ID that is set as active.

1.3.3.18   Command CMD_SLAVE_GET_VALUE (0x17)

The CMD_SLAVE_GET_VALUE function reads the measured value from the sensor. The function has a parameter (FLAG), which determines the method of reading the value from the sensor, Table 1.37. The primary flags that must be implemented are FLAG_DATA_SINGLE, FLAG_DATA_COEFICIENT, and FLAG_DATA_MEMORY. The implementation of other flags depends on the type of measuring module. The uBUS protocol specification says that each internal submodule contains a sensor for measuring one physical quantity. The exception is "communication bus bridge" modules, which can connect several sensors for a given submodule. An example is the OWB hardware module. Additional flags are defined for this module, which can be used to access the sensors and their measured values. The CMD_SLAVE_GET_VALUE function has no data part in the request unless otherwise defined.

Table 1.37 Specification of the data frame for the command CMD_SLAVE_GET_VALUE

Action/byte	0	1	2	3	4	5	6	7
Request	0xNN	0xNN	0x17	FLAG				
Response OK	0xNN	0xNN	0x17	DATA (variable length)				

FLAG = 0x01 (FLAG_DATA_SINGLE)
The response will contain a value from one measurement. This functionality must be implemented for each submodule that contains a sensor. Error responses:

1. ERR_NO_VALUE—0x22. No value is prepared
2. ERR_IO_OUTPUT—0x26. Error writing data: device is input.

FLAG = 0x02 (FLAG_DATA_ACC)
Accumulated value from a series of measurements. The average value from these measurements is calculated. This function may not be implemented.

FLAG = 0x50 (FLAG_DATA_COEFICIENT)
Each sensor can have 3 parameters defined, which can be used for later calculation of the final value. The function reads and returns the internal coefficient of the sensor. The number of coefficients is max. 3, numbered from the value 1. This function contains a data part with a length of 1 byte in the request. The number of coefficients is indicated in the request at position 4.
Response format:

- response[3]—the numeric format of the required coefficient. The format is described in Table 1.7.
- response[4..7]—the value of the required coefficient. This data is read from the sub-slave configuration (Fig. 1.21).

Error codes:

- ERR_FLAG_DATA_FORMAT—0x25. Parameter index was not specified
- ILLEGAL_DATA_VALUE—0x03. Parameter index was not allowed

FLAG = 0xAF (FLAG_DATA_MEMORY)
Function for access to the configuration memory. Memory is accessed in blocks. There is always one block in the response, i.e., 16 bytes. This function contains a 1-byte data part in the request, and the block number is at position 4 in the request. The allowed page value is from 0 to 63. Error codes:
ILLEGAL_DATA_VALUE (0x03)—memory address/page is invalid
Specific functions for the module OWB

FLAG = 0x20 (FLAG_DATA_1W_SENSOR_ADDRESS)
It returns the ID of the currently selected 1-wire sensor. The answer contains 8 bytes.

FLAG = 0x21 (FLAG_DATA_1W_SENSOR_INDEX)
It returns the ID of the i-th sensor on the 1-wire bus. This function contains a 1-byte data part in the request. The sequence number of the sensor is entered in the request at position 4. The response contains the ID of the requested sensor with a length of 8 bytes.

FLAG = 0x24 (FLAG_DATA_1W_VALUE)
Reading the measured value from the selected sensor on the 1-wire bus. This function includes a 1-byte or 8-byte data part in the request. The sensor can be addressed in 3 ways:

1. By the index of the 1-wire sensor: *request[4]* or by the value in the data part of the request
2. By entering the placeholder index *request[4] = 0xFF*, when the currently selected sensor will be used. Setting the address is done using the subcommand FLAG_DATA_1W_SENSOR_ADDRESS, FLAG_DATA_1W_SENSOR_INDEX
3. By 8 Byte address of 1-wire sensor: *request[4..11]*

The length of the response depends on the sensors used. When using the DS18B20 digital thermometers, the response contains 9 bytes. The last, ninth byte is the DATA_READY flag, which tells about the validity of the data: 1—the data in the response is valid and 0—the data is not valid. This condition can occur when multiple readings are attempted without a request to start a new measurement. Error codes:

- DEVICE_BUSY—0x06. Measurement on the sensors is in progress.
- ILLEGAL_DEVICE_ADDRESS—0x02. The addressed device does not exist on the bus.
- ILLEGAL_DATA_VALUE—0x03. The 1-w sensor index was not defined.

FLAG = 0x25 (FLAG_DATA_1W_FIRST)
Command related to addressing or scanning the 1-wire bus. The command is used to send the address of the first device on the 1-wire bus. The order of the sensors on the 1-wire bus is given according to the scanning algorithm (Sect. 1.4.1—1-Wire Application Protocol). The arrangement of the sensors on the 1-wire bus is fixed according to their identifiers. The answer contains:

- 8-byte identifier of the 1-wire sensor (response[3–10])
- Flag NEXT (response[11])—information on whether there is another device/sensor on the 1-wire bus. The NEXT flag has the value 1 if there are minimum of two 1-wire sensors.

FLAG = 0x26 (FLAG_DATA_1W_NEXT)
Loading another device on the 1-wire bus. The meaning of the abbreviations is the same as for FLAG_DATA_1W_FIRST. This command can be used repeatedly as long as there is no NEXT=0 flag in the response.

1.3.3.19 Command CMD_SECURE_INIT (0xE1)

Initialization of the communication encryption layer. The communicating master device (superior computer) sends a 0xE1 request to the measuring module with the public key of the master node. The second communicating party—the slave module—sends a response containing its public key (Table 1.38). Public keys are used to encrypt the message for the communicating partner.
Notes:

- keyREMOTE—generated by the communicating peer or the parent computer. This key is used to encrypt packets sent to communicating partners.

Table 1.38 Specification of the data frame for the command CMD_SECURE_INIT

Action/byte	0	1	2	3..5	6..8
Request	0xNN	0xNN	0xE1	keyREMOTE.n	keyREMOTE.e
Response OK	0xNN	0xNN	0xE1	keyLOCAL.n	keyLOCAL.e

Table 1.39 Specification of the data frame for the command CMD_SECURE_ENCODE_KEY

Action/byte	0	1	2	3..5	6..8
Request	0xNN	0xNN	0xE2		
Response OK	0xNN	0xNN	0xE2	keyLOCAL.n	keyLOCAL.e

- keyLOCAL—public key generated/selected by the measuring device. This key is used to encrypt packets sent to the measuring device.

1.3.3.20   Command CMD_SECURE_ENCODE_KEY (0xE2)

The command request to send the public key of the measuring device. The response contains the slave module's new public key, which will be used to encrypt messages sent to this slave module (Table 1.39).

1.3.3.21   Broadcast Communication

In broadcast communication, each MultiSlave must receive and process a message. No response is returned in this mode. Functions in the broadcast mode are:

- RESET (0x01)—reset of the entire MultiSlave device
- START (0x20)—start for the mode of cyclic reading off values from the connected sensor.
- STOP (0x21)—termination of the mode of cyclic reading of values from sensors

1.4   One-Wire Communication Protocol

The 1-wire protocol is a single-wire interface, half-duplex, bidirectional, low-speed and power, long-distance serial-data communication protocol. The 1-wire protocol is used in the presented sensor system as the primary way to implement the temperature sensors in situations where more sensors are needed.

The basis of 1-wireő technology is a serial protocol using a single data line plus a ground reference for communication. A 1-wire master initiates and controls the communication with one or more 1-wire slave devices on the 1-wire bus (Fig. 1.22). Each 1-wire slave device has a unique, unalterable, factory-programmed, 64-bit

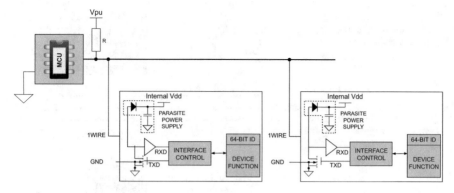

Fig. 1.22 Connecting devices to the 1-wire bus [15]

identification number (ID), which serves as a device address on the 1-wire bus. The 8-bit family code, a subset of the 64-bit ID, identifies the device type and functionality. The following 48 bits (6 bytes) serve as globally unique addresses; the last eight bits are the CRC of the first 56 bits. The 1-wire protocol uses CMOS/TTL logic and operates at a supply voltage range of 2.8 to 6 V. Master and slave can be receivers and transmitters, but only one direction at a time. The LSB always goes first. Time slots transfer data in the 1-wire network [15].

The 1-wire is a voltage-based digital system with two pins, data and ground, for half-duplex bidirectional communication. Compared to other serial communication systems such as I^2C or SPI, the 1-wire devices are designed for use in a momentary contact environment. Disconnecting from the 1-wire bus or losing contact puts the 1-wire slaves into a defined reset state. When the voltage returns, the slaves wake up and signal their presence. With only one contact to protect, the built-in ESD protection of 1-wire devices is exceptionally high. With two contacts, 1-wire devices are the most economical way to add electronic functionality to nonelectronic objects for identification, authentication, and delivery of calibration data or manufacturing information.

Slaves can be powered in two modes (Fig. 1.23):

- Externally Powered: A power pin on the slave device is used to power up the slave on the bus. This topology is used when a slave has high power needs.
- Parasitically Powered: The slave is powered by the data line. Slave devices have an internal capacitor that stores this energy when the bus is idle and pulled up by the weak pull-up.

The following are the features of 1-wire interface protocol:

- It uses a single data line, and no clock is needed. At least two wires (i.e., data and GND) are used for the 1-wire protocol.
- The clock signal is not required as slave devices use the internal clock. This internal clock signal is synchronized with the signal from the master device.
- It has a half-duplex communication mechanism.

Fig. 1.23 1-wire powering mode—top: normal mode and bottom: parasite mode

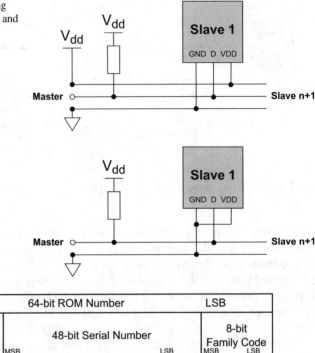

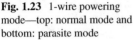

MSB	64-bit ROM Number		LSB
8-bit CRC	48-bit Serial Number		8-bit Family Code
MSB LSB	MSB	LSB	MSB LSB

Fig. 1.24 64-bit unique ROM number

- It has less hardware complexity but more software complexity in implementing the algorithm.
- Due to less wiring, it is a cheaper interface protocol, and hence it is economical.
- It uses a supply voltage between 2.8 V and 5.25 V.
- It uses a 64-bit device addressing scheme—Fig. 1.24.
- The multiple slaves are supported in a multi-drop mode.
- It supports a data rate of 16.3 kbps in standard mode and 163 kbps in overdrive mode.
- It has less power consumption.

Network Topology When designing a 1-wire sensor network, two parameters that affect bus limits are important: radius and weight.

- The radius of the network is the distance between the master node and the farthest slave node measured in meters.
- The weight of the network equals the total sum of active bus lengths measured in meters. Those are the bus parts to which the slave nodes are connected.

For example, in the case of a star topology, where individual branches have lengths of 5 m, 15 m, and 20 m, the diameter is 20 m (the most extended branch), and the

weight is 40 m (5 m + 15 m + 20 m). In general, the network's weight limits the
signal's rise time on the conductor. At the same time, the radius determines the time
of the slowest signal reflection in the communication [16].

Each connected slave node contributes a specific value to the total weight of the
bus—this property is referred to as the weight of the slave device. According to [16],
the weight of the slave node, which is of an iButton type, is 1 m. Other devices have
a weight of 0.5 m. The following topologies are recommended when designing a
1-wire bus topology:

1. Linear topology—a bus is a simple pair of wires that starts at the master device
   and ends at the last slave device. The slave devices are connected directly to the
   bus, while the length of individual connections is max. 3 m.
2. Stubbed topology—it is similar to linear topology, but the length of the branches
   is longer than 3 m.
3. Star topology—the slave devices are connected to a single point, and this
   common point is connected to the master device. The length of individual
   branches in the connection can be different. Slave devices can be connected along
   or at the end of individual branches.

Individual supported topologies are shown in Fig. 1.25.

From the point of view of communication reliability, the star topology is the
most problematic. The junction of various branches presents highly mismatched
impedances; reflections from the end of one branch can travel distances equal to
nearly the weight of the network (rather than the radius) and cause data errors. For
this reason, the unswitched star topology is not recommended, and no guarantees
can be made about its performance. The most reliable topology is the linear
topology. Interferences caused by the signal reflected from the end of the bus are
eliminated, thanks to the short length of the individual branches.

Communication The 1-wire protocol is an open collector bus controlled by one
controller that initiates (most) transmissions. The remote device shorts the 1-wire
bus to ground for 480 µs to signal the controller to search for new devices. While
this is not required when devices are permanently connected to a 1-wire bus, the
"presence pulse," as it is called, can still be monitored. All the devices do this when
they first receive power from the 1-wire bus and when a controller asserts a reset
pulse. The open collector bus is typically pulled up to VCC (minimum of 3.0 V)
with a 4.7 kΩ resistor. The bus is considered idle when no device (controller or
remote) is pulling the bus to the ground, and therefore, it will be at a logic 1 state
at or near VCC. When any device pulls the bus to the ground, it will be in a logic
0 state. During the idle time, all remotes will see VCC and an internal diode allows
an internal capacitor to be charged up to the VCC. This internal diode/capacitor
creates a parasitic power supply to allow the device to operate even when the 1-
wire bus is pulled to the ground. The tiny power requirements of remote devices
allow them to remain powered as long as the bus returns to idle within a reasonable
time. The transmission protocol timing always assures that this will happen [17].
A microprocessor can generate 1-wire timing signals if a dedicated bus master

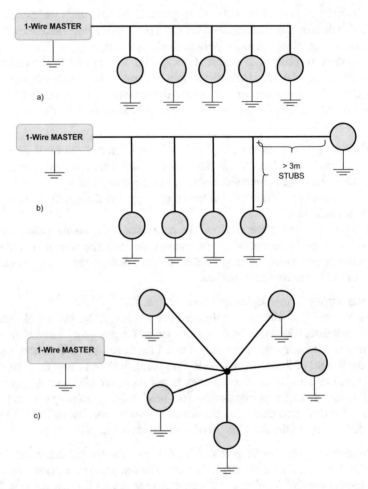

Fig. 1.25 1-wire bus topologies: (**a**) linear, (**b**) stubbed, and (**c**) star

is not present. There are several system requirements for the proper operation of successful communication:

1. The communication port must be bidirectional, its output is open-drain, and there is a weak pull-up on the line. This is a requirement of any 1-wire bus.
2. The system must generate an accurate and repeatable 1 μs delay for standard speed and 0.25 μs delay for overdrive speed.
3. The communication operations must not be interrupted while being generated.

The four basic operations of a 1-wire bus are Reset, Write 1 bit, Write 0 bit, and Read bit. The time it takes to perform one bit of communication is called a time slot in the device data sheets. Byte functions can then be derived from the multiple calls to the bit operations. See Table 1.40 below for a brief description of each operation

Table 1.40 1-Wire Operations [18]

Operation	Description	Implementation
Write 1 bit	Send a "1" bit to the 1-wire slaves (Write 1-time slot)	Drive bus low, delay T1 Release bus, delay T2
Write 0 bit	Send a "0" bit to the 1-wire slaves (Write 0-time slot)	Drive bus low, delay T3 Release bus, delay T4
Read bit	Read a bit from the 1-wire slaves (Read time slot)	Drive bus low, delay T1 Release bus, delay T5 Sample bus to read a bit from the slave Delay T6
Reset	Reset the 1-wire bus slave devices and ready them for a command	Delay T7 Drive bus low, delay T8 Release bus, delay T9 Sample bus, 0 = device(s) present, 1 = no device present Delay 10

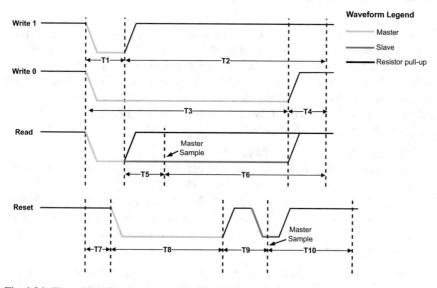

Fig. 1.26 Time slot for 1-wire communication [18]

and a list of the steps necessary to generate it. Figure 1.26 illustrates the waveforms graphically. Table 1.41 shows the recommended timings for the 1-wire master to communicate with 1-wire devices over the most common line conditions. Alternate values can be used when restricting the 1-wire master to a particular set of devices and line conditions. See the downloadable worksheet to enter system and device parameters to determine the minimum and maximum values.

Table 1.41 1-Wire Master Timing [18]

Parameter	Recommended time [ms] standard/overdrive speed
T1—WRITE_ONE	6 / 1.0
T2—WRITE_ONE_DELAY	64 / 7.5
T3—WRITE_ZERO	60 / 7.5
T4—WRITE_ZERO_DELAY	10 / 2.5
T5—RELEASE_DELAY	9 / 1.0
T6—READ_DELAY	55 / 7
T7—START_=DELAY	0 / 2.5
T8—RESET_PULSE	480 / 70
T9—READ_PRESENCE	70 / 8.5
T10—RESET_DELAY	410 / 40

1.4.1 1-Wire Application Protocol

We will separately describe the 1-wire application protocol using a state diagram to communicate master and slave nodes. Slave nodes exist as standardized sensors. They have a communication protocol implemented in their firmware. In the next part, we discuss the implementation of the slave node of the 1-wire bus, and we present the state diagram and implementation of the communication protocol for the slave as well. The 1-wire application protocol defines a basic set of commands that every 1-wire device must implement. They are [19]:

1. Commands for addressing—Table 1.42
2. Commands for implementing the required functionality—they can be commands for setting the configuration (WRITE_SCRATCH, READ_SCRATCH), commands for writing/reading from/to the internal memory (WRITE_MEM, READ_ROM) or starting the temperature conversion for temperature sensors (CONVERT_T). The numerical codes of these functions may change for different slave nodes.

Communication between the master and a slave 1-wire can be in unicast or broadcast mode. The desired mode depends on the first command after the RESET pulse. Figure 1.27 demonstrates the unicast communication. The reset pulse (RESET) is followed by the MATCH_ROM command (MATCH), followed by the address of the desired slave node. The family code of the connected sensor (FAMILY) is displayed first. This code is 0x28, so it is a DS18B20 digital thermometer. Next, it is followed by the sensor ID and the CRC8 checksum. Subsequently, the master sends the required command for the addressed slave node. In this case, it is the READ_SCRATCHPAD (0xBE) command. The slave responds with 3 bytes (0x6A, 0x5D, 0x55) representing the requested data—in this case, it is the temperature from the DS18B20 temperature sensor.

Broadcast communication is shown in Fig. 1.28, where the SKIP_ROM is sent as the first command after the RESET pulse, followed by the command code (0x4E—

Table 1.42 Common 1-wire commands

1-wire command	Value	Description
READ_ROM	0x33	Read a device's ID. Used when there is only one device on the bus
SKIP_ROM	0xCC	Ignore device ID(s). It is intended to be used if only one device is on the bus. However, it can also be used to send subsequent commands to all devices—broadcast mode
SEARCH_ROM	0xF0	Begin enumerating IDs
MATCH_ROM	0x55	Select a device with a specific ID. The following 64-bits to be written will be the known ID

Fig. 1.27 Communication in the unicast mode

Fig. 1.28 Communication in broadcast mode

WRITE_SCRATCHPAD), which is executed on all the connected slave nodes. In this case, it is about sending the setting to all the connected DS18B20 sensors—setting the resolution for the subsequent temperature measurement. 3 bytes are sent as the configuration data: 0x00, 0x00, 0x1F.

The specific command is SEARCH_ROM, belonging to the broadcast group. It is used to search for all the slaves on the 1-wire bus. The "1-wire search algorithm" [20] is used to search the entire bus. The search algorithm is a binary tree search where branches are followed until a device ROM number, or a leaf, is found. Subsequent searches then take the other branch paths until all of the leaves present are discovered. The search algorithm begins with the devices on the 1-wire being reset using the reset and presence pulse sequence. If this is successful, then the 1-byte search command is sent. The search command readies the 1-wire devices to begin the search. Following the SEARCH_ROM command, the actual search begins with all the participating devices simultaneously sending the first bit (least significant) in their ROM number (see Fig. 1.24) As with all 1-wire communication, the 1-wire master starts every bit, whether it is data to be read or written to the slave devices. Due to the 1-wire characteristics, if all devices respond simultaneously, this results in a logical AND of the bits sent. After the devices send the first bit of their ROM number, the master initiates the next bit, and the devices then send the

Table 1.43 1-wire search algorithm state

Bit (true)	Bit (complement)	Information known
0	0	There are both 0s and 1s in the current bit position of the participating ROM numbers. This is a discrepancy
0	1	There are only 0s in the bit of the participating ROM numbers
1	0	There are only 1s in the bit of the participating ROM numbers
1	1	No devices participated in the search

complement of the first bit. From those two bits, information can be derived about the first bit in the ROM numbers of the participating devices (see Table 1.43).

According to the search algorithm, the 1-wire master must send a bit back to the participating devices. If the participating device has that bit value, it continues participating. If it does not have the bit value, it goes into a wait state until the next 1-wire reset is detected. This "read two bits" and "write one bit" pattern is then repeated for the remaining 63 bits of the ROM number. At the end of one pass, the ROM number of this last device is known. On subsequent passes of the search, a different path (or branch) is taken to find the other device's ROM numbers. This algorithm is described in detail in [20].

1.4.2   1-Wire Master

The master node initiates all communication (Fig. 1.29). From the point of view of the master node, communication has several stages—detection of the slave nodes on the bus, searching for devices on the bus or addressing a slave node and sending a specific command. The 1-wire bus master node can be in the states defined in Fig. 1.30. The description of individual states and transitions between them is given in the following text:

1. IDLE—master is idle.
2. DISCOVERY—the master examines/addresses the 1-wire bus.

   a. Master sends a RESET pulse. Existing slave nodes detect reset pulses and prepare for the subsequent communication—Fig. 1.29 time stamp A-B.
   b. After the RESET pulse ends, all the existing slave nodes signal their presence by changing the log bus levels to 0.
   c. The master detects the presence of one or more slave devices after the READ_PRESENCE time has elapsed—Fig. 1.29 time stamp C.
   d. The master sends the command READ_ROM (only if there is precisely 1 slave), MATCH_ROM (searching the 1-wire bus), SEARCH_ROM (verifying

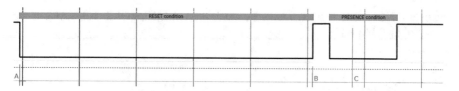

Fig. 1.29 Initialization of 1-wire communication

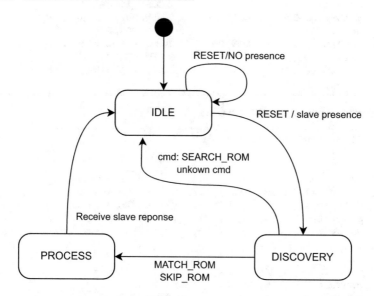

Fig. 1.30 1-wire master state diagram [21]

the existence of a slave with an exact address), or SKIP_ROM (skipping addressing—a broadcast message follows).

3. PROCESS—the master sends the function code to the slave node. In the case of the SEARCH_ROM command, continue with step 1.

a. The addressed slave processes the received code performs the requested function and sends the data. The master node receives this data.

The program implementation of the master node can be realized on any microcontroller, on which one can control the state of the output GPIO pin with a time resolution of 1 us. The implementation of the basic operations is then simple. Assume that the 1-wire bus is connected to the GPIO pin labeled PIN_OW. The principle code for generating the RESET pulse is given in Listing 1.2, the code for writing 1 bit in Listing 1.3, and the code for reading 1 bit in Listing 1.4. The specific definition of the pin address and the delay function depend on the hardware implementation. All the listed function names are prefixed with "*Owm*," which stands for one-wire master. Times T, mentioned in Listings 1.2, 1.3, and 1.4, are defined in Table 1.41.

```
int OwmReset(void)
{
    int result;
    delay(T7);                        // START_DELAY
    pinWrite(PIN_OW, 0x00);           // Drives DQ low
    delay(T8);                        // RESET_PULSE
    pinWrite(PIN_OW, 0x01);           // Releases the bus
    delay(T9);                        // READ_PRESENCE
    result = pinRead(PIN_OW);         // Sample for presence pulse
    delay(T10);                       // RESET_DELAY
    return result;                    // sample presence result
}
```

Listing 1.2 Implementation of reset pulse

```
void OwmWriteBit(int bit)
{
  if (bit)
  { // Write '1' bit
    pinWrite(PIN_OW, 0x00);   // Drives DQ low
    delay(T1);                // WRITE_ONE
    pinWrite(PIN_OW, 0x01);   // Releases the bus
    delay(T2);       // WRITE_ONE_DELAY, Complete the time slot
  }
  else
  {                           // Write '0' bit
    pinWrite(PIN_OW, 0x00);   // Drives DQ low
    delay(T3);                // WRITE_ZERO
    pinWrite(PIN_OW, 0x01);   // Releases the bus
    delay(T4);                // WRITE_ZERO_DELAY
  }
}
```

Listing 1.3 Implementation of 1-bit writing

```
int OwmReadBit(void)
{
  pinWrite(PIN_OW, 0x00);   // Drives DQ low
  delay(T1);                // WRITE_ONE
  pinWrite(PIN_OW, 0x01);   // Releases the bus
  delay(T5);                // RELEASE_DELAY
  int result = pinRead(PIN_OW); // Sample the bit value
  delay(T6);       // READ_DELAY, Complete the time slot
  return result;
}
```

Listing 1.4 Implementation of reading 1 bit

1.4.3   1-Wire Slave

Each slave device of the 1-wire bus must contain a unique ID, which consists of a family code (8-bit), the ID itself (48-bit), and a checksum (8-bit)—Fig. 1.24. This unique number is stored in the internal ROM memory of each slave. In addition, each slave node contains internal rewritable memory. It can be SRAM or EEPROM, depending on the specific type. A buffer known as Scratchpad is used to access this memory. Its size depends on the particular slave. The 1-wire slave may be in one of the three states (Fig. 1.31):

- IDLE
- Function CMD
- Control CMD

The 1-wire slave does not perform any activity in the IDLE state and is waiting for incoming communication. In the function CMD state, a command related to addressing is processed—searching algorithm (SEARCH_ROM), address check (MATCH_ROM), or transition to broadcast mode (SKIP_ROM). In the control CMD state, the slave performs the requested command. According to conventions, the CONVERT command is issued to start the measurement, READ_SCRATCHPAD command to read the SCRATCHPAD area, which stores the measured value, and READ_MEMORY command to read the additional memory contents. The slave in the IDLE state is in a waiting loop for the RESET pulse. If an active value appears on the bus, the slave will start detecting the RESET pulse. In case the RESET pulse is too short or too long, the slave remains in the IDLE state. If the RESET pulse is evaluated as correct, the slave switches to function CMD (Fig. 1.31). The RESET pulse length is 480µs in the standard mode. After the RESET pulse, the bus returns to the inactive state for 70µs. After that time, all the connected slaves signal their presence on the bus by changing the bus state to logical 0. If the master detects the logical 0 at this time, it means that there is at least 1 slave connected on the bus, and the operation is followed by an enumeration of connected slaves using the "1-wire search algorithm." Upon completing this algorithm, the master has received all the slave identifiers on the 1-wire bus. Using that information, the master can address a specific slave and request measured data from the sensor [21].

For the 1-wire slave, flow charts were defined (Figs. 1.32 and 1.33) describing the management of the requests. Those diagrams define the behavior of the 1-wire slave. Figure 1.32 shows a flowchart for processing commands in the CMD state function. The READ_ROM command will request the slave address. Subsequently, the slave sends its address to the 1-wire bus. That command is used to verify that the slave is still present on the bus. The MATCH_ROM command is processed as follows: after issuing the command, the master sends the individual bits of the slave node address to be addressed. If all the sent bits match those of the slave address, the slave changes its status to the function command; otherwise, its status returns to the default IDLE state. The SEARCH_ROM command triggers the bus searching algorithm. The last command is SKIP_ROM, which the slave transfers into the function CMD state

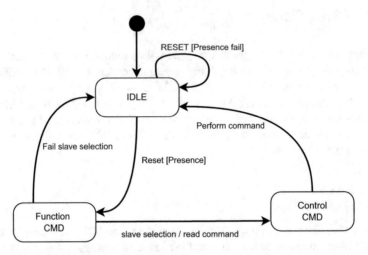

Fig. 1.31 1-wire slave state diagram [21]

without checking the address. This results in a situation where all the connected slave nodes will execute the sent command. Switching the bus into a broadcast state is possible when all the connected 1-wire slaves are active [21].

Figure 1.33 shows the command processing algorithm in the function CMD state. The first command is CONVERT, which starts the measurement on the connected slave.

Implementation of the 1-Wire Slave Module When implementing a 1-wire slave device, it is necessary to implement elementary functions for processing the RESET pulse, reading, and writing a bit. According to the diagram in Fig. 1.32, it is further needed to implement the SEARCH ROM and MATCH ROM functions. Other functions can be quickly implemented using a combination of already existing functions. In the following source codes, the microcontroller pin, by means of which the 1-wire slave is connected to the bus, is marked as OWS_PIN.

Processing of the RESET pulse on the side of the slave device is demonstrated by the *OwsResetPulse* function in Listing 1.5. The function returns the value 1 upon successful detection of the RESET pulse. Otherwise, it returns the value 0.

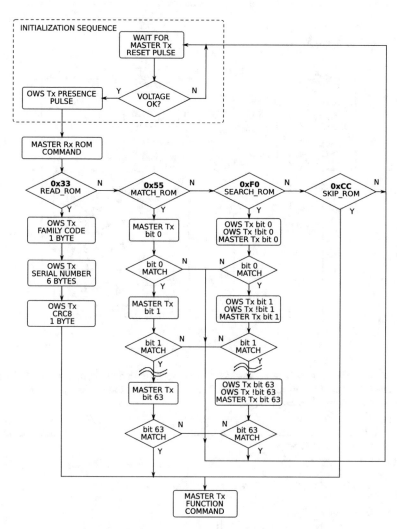

Fig. 1.32 1-wire slave function commands processing algorithm (Copyright © 2022, Maxim Integrated Products Inc., all rights reserved. Maxim Integrated Products Inc. is a wholly owned subsidiary of Analog Devices Inc.) [22]

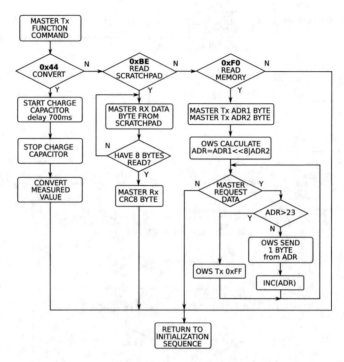

Fig. 1.33 1-wire slave control commands processing algorithm (Copyright © 2022, Maxim Integrated Products Inc., all rights reserved. Maxim Integrated Products Inc. is a wholly owned subsidiary of Analog Devices Inc.) [22]

```
int OwsResetPulse(void)
{
  int count=0;
  while(count<50)  // 50*10us => 500us max
  {
    count++;
    delay(10);   // delay for 10us
    if(pinRead(PIN_OWS) == 1)
    { // end of RESET pulse detection.
      break;
    }
  }
  if(count >= 50)
  { // RESET pulse too long
    return 0;
  }
  if(count < 40)
  { // RESET pulse too short: <400us
    return 0;
  }
  return 1; // valid RESET pulse
}
```

Listing 1.5 Detection of RESET pulse on the slave device

The *OwsWriteBit* function is designed for writing 1 bit—Listing 1.6. Writing 1 bit to the 1-wire bus consists of setting the bus to the value log.0 for a precisely defined time. This time is different for writing the values 0 and 1. The 1 bit writing sequence must always be initialized by the master device, which changes the bus state to the value log.0 for the duration of WRITE_ONE. Subsequently, the slave node must hold the line in the log.0 state (bit 0 written) for OWS_WRITE_DELAY or release the bus to get to the log.1 state (bit 1 written). The master waits for the RELEASE_DELAY time and then reads the level of the 1-wire bus. This situation is shown in Fig. 1.34 from the perspective of both the master and slave nodes. Reading 1 bit by the master was described in the previous section. From the slave device point of view, log.0 is written to the bus after the reading pulse is detected only if the slave device sends the log.0 value. If it sends a value of log.1, the pull-up resistor ensures that the state of the bus changes to the value of log.1. Subsequently, the OWS_READ_DELAY period waits until the reading of the bus status (master sample) takes place on the master device. As the last operation, the bus returns to the idle state (writing the log.1 value) and waits for the end of the time slot.

```
void OwsWriteBit(int data) {
  if(data == 0){
    pinWrite(PIN_OWS, 0);
  }
  delay_us(OWS_WRITE_DELAY);
  pinWrite(PIN_OWS, 0x01);
  delay_us(OWS_SLOT_DELAY);
}
```

Listing 1.6 One-wire slave write 1 bit to the bus

The last elementary operation is reading 1 bit. In the case of this operation, it is necessary to detect the error conditions that might occur. The slave node can start reading a bit on the bus only after the master node changes the bus level state to log.0. The first step in reading a bit is to detect the log.0 state on the bus. In Listing 1.7, the timer marked as *tim1* is used in interrupt mode. The timer has a timeout set to 400 μs. In the beginning, the slave waits until the bus status changes

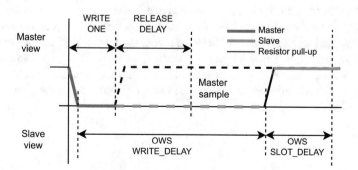

Fig. 1.34 Writing 1 bit by the slave device

to log.0. If the 400 μs time expires and the bus level is still at log.1, the timer operator sets the flag ow_flag_error_read_bit=1, and then the global error flag ow_error is set to the value 1. The subsequent read value of the bus level is therefore invalid. After the log.0 level is detected, the OWS_READ_WAIT period waits, and the slave device reads the logic level of the 1-wire bus (Fig. 1.35). The last thing is another check of the validity of the read value. If the bus level is at the log.0 for a period longer than 50 μs from the sampling time, it is assumed that this is a RESET pulse. The flag of the read value validity is set to 0 again.

```
int OwsReadBit(void) {
  uint8_t data;
  ow_flag_error_read_bit = 0;
  SetTimer(tim1, MODE_INT);
  StartTimer(tim1, 400);
  while(pinRead(PIN_OWS) == 1) {  // detection for write timeout
    if(ow_flag_error_read_bit == 1) {
      ow_error = 1;
      break;
    }
  }
  delay_us(OWS_READ_WAIT);     // delay before read
  data = pinRead(PIN_OWS);     // read the value
  int tmr=0;   // detecting read error - long time for value 0
  while(pinRead(PIN_OWS) == 0) {
    delay(5);
    tmr+=5;
    if(tmr > 50) {
      ow_error = 1;
      return 0;
    }
  }
  return data;
}
```

Listing 1.7 One-wire slave read 1 bit from the b

Each 1-wire bus slave device must implement addressing algorithms, which are implemented using the SEARCH_ROM and MATCH_ROM functions. The MATCH_ROM function must check whether the address sent by the master to the bus matches the own address of the slave module. In Listing 1.8, this same function is implemented. In the *OwsMatchROM* function, the 8 bytes sent by the master are checked sequentially. Those 8 bytes are the slave address. The task of this function is to determine whether this address is identical to the slave node's address. The function's output is information about the match, or mismatch, of the broadcast and its own address.

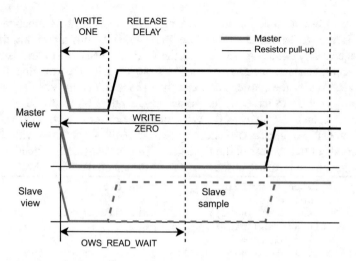

Fig. 1.35 Reading 1 bit by the slave device

```
uint8_t OwsMatchBit (uint8_t read_bit) {
  uint8_t result=0;
  if (OW_read_bit() == read_bit){
    result = 1;              // match
  }
  if(ow_error){
    result = 0;              // not match
  }
  return result;
}

int OwsMatchROM(void) {
  int result=1;
  i = 7;                     // match for 8 bytes
  do {
    current_byte = OneWireSID[i];
    int bit=0;
    while(bit<8){            // 8bit per byte
      while(OW_read_bit() == 1);
      if(!OW_match_bit(current_byte & 0x01)) {
        result=0;
      }
      current_byte >>= 1;
    }
  } while(i--);
  return result;
}
```

Listing 1.8 MATCH_ROM function implementation

The most crucial function in addressing is the MATCH_ROM function, which is used in the bus search algorithm (see Table 1.43). In this algorithm, the slave node sequentially sends its address to the bus: the LSB bit of the address is always sent first. After sending one bit, the slave waits for the initiation of further communication by the master node and then sends the complementary value of the already sent bit to the bus. The master then sends 1 bit to the bus. The slave reads this bit, and if it matches the value of the address_bit bit (Listing 1.9, in function *OwSearchROM*, OW_match_search call), it continues with checking the other bits of the address of the slave node. The *OwSearchROM* function returns information on the presence/absence of a slave device on the 1-wire bus.

```
uint8_t OW_match_search (uint8_t address_bit) {
  uint8_t result=0;
  while(OW_read_bit() == 1);       // wait until bus is inactive
  OwsWriteBit(address_bit);        // send bit
  while(OW_read_bit() == 1);
  OwsWriteBit(address_bit ^ 0x01); // sent complement bit

  // read bit and compare with sent one
  if(OwsReadBit() == address_bit)
    result = 1;                    // match
  if(ow_error){
    result = 0;                    // error during reading 1-wire line
  }
  return result;
}

uint8_t OwSearchROM(void) {
  int i = 7, j;
  do {
    current_byte = OneWireSID[i];
    j=8;
    while(j >= 0){     // sending LS-bit first
      if(!OW_match_search(current_byte & 0x01)) {
        return  0;
      }
      current_byte >>= 1;
      j--;
    }
  }
  return 1;
}
```

Listing 1.9 SEARCH_ROM function implementation

1.4.4   Dynamic Control of the 1-Wire Bus Status

According to the definition of 1-wire network parameters, the essential parameters are the bus's radius and weight. Each slave node contributes an average value of 1 m to the total weight of the bus. According to [16], a simple 1-wire network that uses only pull-up resistors has a maximum weight of 200 m. It is necessary to use an active increase of the bus voltage level to achieve higher values of the weight. This solution will ensure a higher speed of changing the logic level of the bus. The document [23] defines an active pull-up, as shown in Fig. 1.36.

R4, R1, and R3 series paths provide the standard 1-wire pull-up to VCC. With this circuit, the total pull-up resistance is approximately 1 kΩ. This value applies when the 1-wire line is idle. Since R4 is connected to the Q1 drain, the current flowing through it does not affect the low-level voltage on the 1-wire line. Resistors R4, R1, and R3, with the load of the 1-wire network "weight," determine the speed of the voltage on the 1-wire rises to 5V. The Schottky diodes D1 and D2 eliminate spikes from electrostatic discharge (ESD) hits or cross-coupling from

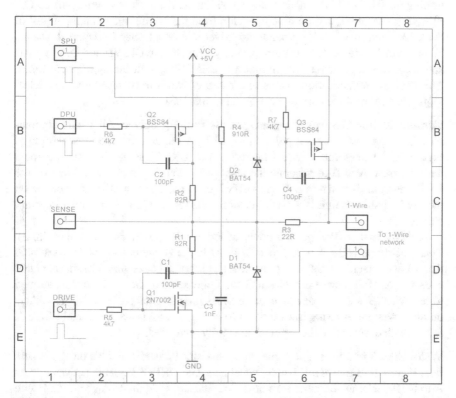

Fig. 1.36 Active pull-up schematic [23]

another cabling nearby by conducting them to GND and VCC, respectively. R3 limits the ESD current and protects D1 and D2.

A unique characteristic of this driver is the proper line termination of the 1-wire cable on the master end. The category 5 unshielded twisted-pair data cable, recommended for 1-wire applications, has a characteristic impedance of approximately 100Ω. Line termination is accomplished through R1 or R2 in series with R3 when Q1 or Q2 is conducting. C3 in series with R1 and R3 provides an AC-coupled termination for presence pulses. R1 and R2 need to be changed accordingly to adapt this driver to a different impedance. All three sections of this driver are slew rate controlled when the associated transistor is switched on. R5 and C1 limit the slew rate when the driver pulls the 1-wire line low, e.g., at the beginning of a time slot or a reset pulse. R6 and C2 limit the slew rate when the active pull-up becomes active. R7 and C4 limit the slew rate of the active pull-up. The time constant of all three sections is 0.5 μs. This value results in a slew rate of approximately $4V/\mu s$.

Disregarding the robust pull-up circuitry (Q3, R7, C4), the driver requires three connections to a supervising microcontroller. These signals are called DRIVE, DPU, and SENSE. DRIVE is an active-high signal that initiates 1-wire communication by turning on Q1. DPU is an active-low signal that activates the dynamic pull-up Q2. SENSE is essentially a through-connection from the 1-wire line to an input port of the microcontroller. 1-wire ground and driver/microcontroller GND are the same. To perform 1-wire communication, properly generate the DRIVE and DPU signals and read from the 1-wire line through the SENSE input at the appropriate times. The 1-wire communication knows four cases of waveforms: reset/presence-detect sequence and the three communication time slots [23].

Communication Control Using a Dynamic Pull-Up Driver All 1-wire communication begins with a reset pulse followed by a window for the presence pulse. Figure 1.37 shows the 1-wire waveform. The DRIVE signal is activated for a duration from A to B to generate the reset pulse. Starting at A, the voltage on the 1-wire bus ramps down to 0 V. As DRIVE becomes inactive at B, the voltage on the 1-wire bus rises unless a 1-wire device pulls the line low to signal an interrupt condition.

After the first DRIVE pulse is done, an active pull-up DPU is activated (point E). This quickly pulls the 1-wire line to 5 V. At F, the active pull-up ends. Assuming that a 1-wire device is present, it generates a presence pulse, which begins at G and ends at I. At H, somewhere between G and I, the status of the 1-wire bus is sampled to test whether a 1-wire device is present. As the presence pulse ends, the voltage on the 1-wire starts rising toward 5V. The active pull-up is activated again from K to L, which ensures that the 1-wire line is fully recharged.

Write Zero Time Slot The write zero time slot transmits a 0 bit on the 1-wire line. Figure 1.38 shows the 1-wire waveform. The DRIVE signal is activated from A to B0 to generate the write zero time slot. Starting at A, the voltage on the 1-wire ramps down to 0 V. As DRIVE becomes inactive at B0, the voltage on the 1-wire line rises. Shortly after B0, the active pull-up DPU is activated (point C0). This

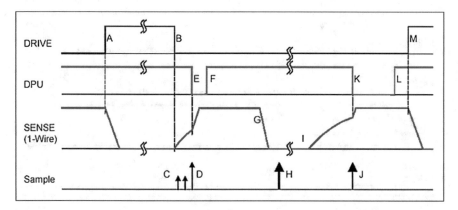

Fig. 1.37 Reset and presence-detect sequence (Copyright © 2022, Maxim Integrated Products Inc., all rights reserved. Maxim Integrated Products Inc. is a wholly owned subsidiary of Analog Devices Inc.) [23]

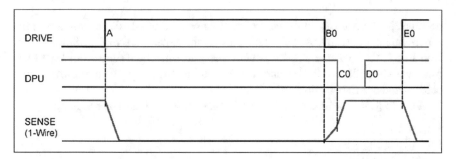

Fig. 1.38 Write zero time slot (Copyright © 2022, Maxim Integrated Products Inc., all rights reserved. Maxim Integrated Products Inc. is a wholly owned subsidiary of Analog Devices Inc.) [23]

quickly pulls the 1-wire line to 5 V. At D0, the active pull-up ends. The next time slot or a reset/presence-detect sequence may follow at E0.

Refer to [23], recommended Timing Values for Write Zero Time Slot are:

- A to B0 : 60 μs
- B0 to C0 : 2 μs
- C0 to D0 : 16 μs
- A to E0 : 80 μs

Write One/Read Time Slot (Read One) The write one time slot transmits a 1-bit on the 1-wire line. Figure 1.39 shows the 1-wire waveform. Reading a 1-bit from the 1-wire line results in exactly the same waveform as writing a 1-bit. Therefore, the write one and read one case are combined and discussed as a single case. To generate the write one or read time slot, the DRIVE signal is activated for a duration from A to BR. Starting at A, the voltage on the 1-wire ramps down to 0V. As DRIVE

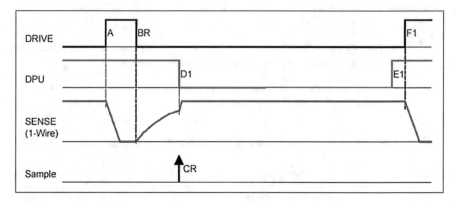

Fig. 1.39 Write one/read time slot (Copyright © 2022, Maxim Integrated Products Inc., all rights reserved. Maxim Integrated Products Inc. is a wholly owned subsidary of Analog Devices Inc.) [23]

becomes inactive at BR, in the case of reading or writing a 1-bit, the voltage on the 1-wire line starts rising. At CR, the status on the 1-wire is sampled. Since the bit read is a 1, the dynamic pull-up is activated immediately, lasting from D1 to E1. This quickly pulls the 1-wire line to 5 V. The next time slot or a reset/presence-detect sequence may begin at F1.

Refer to [23], recommended Timing Values for Write One Time Slot are:

- A to BR : 9 μs
- A to CR : 18 μs
- CR to D1 : $0 to 2$ μs
- E1 to E1 : 60 μs
- A to F1 : 80 μs

Read Time Slot (Read Zero)   The read zero time slot is a read time slot that reads a 0 bit from the 1-wire line. Figure 1.40 shows the 1-wire waveform. To generate the read time slot, the DRIVE signal is activated for a duration from A to BR. Starting at A, the voltage on the 1-wire ramps down to 0V. To send a 0 bit, a 1-wire device starts pulling the 1-wire line low after A, but before BR. The voltage on the 1-wire line, therefore, is first driven low by the 1-wire master and then remains held at a logic low by one or more 1-wire devices. When BR has occurred, the master pulldown is turned off. At CR, the status on the 1-wire is sampled. Since the bit read is a 0, the 1-wire is sampled repeatedly. At D0, the 1-wire device stops pulling the line low, which allows the voltage to rise. A subsequent sampling at E0 determines that the line has reached a logic-high level. Now the dynamic pull-up is activated, lasting from F0 to G0. This quickly pulls the 1-wire line to 5 V. The next time slot or a reset/presence-detect sequence may begin at H0.

Recommended Timing Values for Read One Time Slot are:

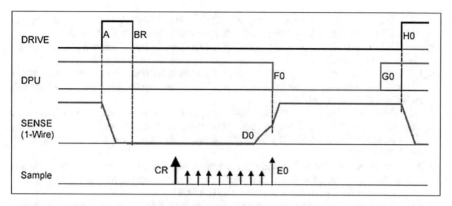

Fig. 1.40 Read time slot (Copyright © 2022, Maxim Integrated Products Inc., all rights reserved. Maxim Integrated Products Inc. is a wholly owned subsidiary of Analog Devices Inc.) [23]

- A to BR : 9 µs
- A to CR : 18 µs
- E0 to F0 : $0\,to\,2$ µs
- CR to G0 : 60 µs
- A to H0 : 82 µs

The presented dynamic pull-up solution was the basis for the design of the OWB hardware module, described in Sect. 2.1. The OWB hardware module is a 1-wire bus driver with 1-wire bus length detection and automatic setting of the bus timing parameters.

References

1. J. Ďuďák, G. Gašpar, S. Sedivy, P. Fabo, L. Pepucha, P. Tanuska, Serial communication protocol with enhanced properties–securing communication layer for smart sensors applications. IEEE Sensors J. **19**(1), 378–390 (2019). https://doi.org/10.1109/JSEN.2018.2874898
2. Telecommunications Industry Association, TIA/EIA-485-A STANDARD. https://www.mikrocontroller.net/attachment/428561/eia485.pdf, 1998 (accessed September 1, 2022). Electrical Characteristics of Generators and Receivers for Use in Balanced Digital Multipoint Systems. Accessed: 5.6.2022
3. Maxim Integrated, APP 763 - Guidelines for Proper Wiring of an RS-485 (TIA/EIA-485-A) Network. https://pdfserv.maximintegrated.com/en/an/AN763.pdf, 2001 (accessed July 15, 2022). Accessed: 2022-06-06
4. Contemporary Control Systems, Inc., Understanding EIA-485 Networks. https://www.ccontrols.com/pdf/ExtV1N1.pdf, 1999, Vol. 1(3) (Accessed September 7, 2022)
5. modbus.org, Modbus over serial line specification and implementation guide v1.02. Technical report, Organization MODBUS, 12 2006
6. S. Mackay, E. Wright, D. Reynders, J. Park, EIA-232 overview, in *Practical Industrial Data Networks*, ed. by S. Mackay, E. Wright, D. Reynders, J. Park (Newnes, Oxford, 2004), pp. 32–52. ISBN: 978-0-7506-5807-2. https://doi.org/10.1016/B978-075065807-2/50027-3. https://www.sciencedirect.com/science/article/pii/B9780750658072500273. Accessed: 2022-03-04

7. B&B Electronics Manufacturing Company, RS-422 and RS-485 Application Note. https://www.cpii.com/docs/library/4/485appnote.pdf, June 2006. Accessed: 2022-04-16

8. J. Ďuďák, *Príspevok k priemyselným komunikačným štandardom.* Disertation thesis, Slovak University of Technology, 2011

9. J. Ďuďák, M. Skovajsa, I. Sladek, Proposal of a communication protocol for smart sensory systems, in *Proceedings of the 16th International Conference on Mechatronics - Mechatronika 2014* (Brno, CZ, 2014), pp. 107–112

10. M. Skovajsa, P. Fabo, L. Pepucha, I. Sládek, Proposal of a wireless measurement system for temperature monitoring of biological active materials, in *Advanced Mechatronics Solutions*, ed. by R. Jablonski, T. Brezina (Springer International Publishing, 2016), pp. 367–372. ISBN: 978-3-319-23923-1

11. ARM Developer, Cortex-M0 technical reference manual: Instruction set summary. Technical report, Arm Limited, May 2021. http://infocenter.arm.com/help/index.jsp?topic=/com.arm.doc.ddi0432c/CHDCICDF.html. Accessed: 2022-11-08

12. ARM Developer, Cortex-M0+ technical reference manual: Instruction set summary. Technical report, Arm Limited, May 2021. http://infocenter.arm.com/help/index.jsp?topic=/com.arm.doc.ddi0432c/CHDCICDF.html. Accessed: 2022-11-08

13. Q.D. Elaine Barker, Recommendation for key management. NIST Special Publication Revision 1, National Institute of Standards and Technology, January 2015. Part 3: Application-Specific Key Management Guidance

14. M. Gracov, Use of microcontrollers in embedded applications. Master thesis, Slovak University of Technology, 2019,

15. Bernhard Linke, Overview of 1-Wire Technology and Its Use. Tutorial, June 2018. https://www.analog.com/en/technical-articles/guide-to-1wire-communication.html. Accessed: 2022-03-08

16. Maxim Integrated, Guidelines for reliable long line 1-wire networks. Tutorial 148, Maxim Integrated Products, Inc., September 2008

17. J. Bachiochi, Putting 1-wire protocol into action. Magazine article Issue 373, CIRCUIT CELLAR MAGAZINE, August 2021

18. Maxim Integrated, 1-wire communication through software. Application Note 126, Maxim Integrated Products, Inc., May 2002

19. Maxim Integrated, WHITE PAPER 5: USING 1-WIRE APIS FOR DATA SHEET COMMANDS. Application Note 1100, Maxim Integrated Products, Inc., June 2002

20. Maxim Integrated, 1-wire search algorithm. Application Note 187, Maxim Integrated Products, Inc., March 2002

21. J. Ďuďák, P. Tanuska, G. Gabriel, P. Fabo, Arm-based universal 1-wire module solution. J. Sensors Article ID 5268247(1), 16 p. (2018). https://doi.org/10.1155/2018/5268247

22. Maxim Integrated, Programmable resolution 1-wire digital thermometer. Datasheet DS18B20, Maxim Integrated Products, Inc., March 2007

23. Bernhard Linke, Advanced 1-wire network driver. Reference design 244, Maxim Integrated Products, Inc., May 2003. Accessed: 2022-10-03

Chapter 2
Hardware Modules of the Sensory System

Hardware and software - engineered to work together.

When designing a sensory system, which is a part of the control system of the selected technological process, it is necessary to prioritize the requirements for the control system parameters and those resulting parameters imposed on the sensory system. We can choose the basic parameters: compatibility with communication buses and communication protocols used by the control system, the speed of the system's response to requests, and the solution's modularity. The single sensors, connected to the measuring units, have their parameters determined individually based on the requirements for solving the given problem.

In the proposed system, the measuring modules represent the fundamental elements of the hardware solution built on the platform mentioned above: to read values from sensors, their (pre)processing, and communication with the measuring center. The uBUS protocol, based on the MODBUS protocol, was chosen as the communication protocol, and the RS-485 bus was chosen as the primary communication platform. The RS-485 bus is often used in industrial environments to communicate between devices with multiple protocols. When designing a separate sensory system intended primarily for data collection and eventual evaluation, we proceed analogously, as in the previous case, from the parameters imposed on the sensory system. Additional parameters can be, for example, the possibility of maintenance, reconfiguration, and diagnostics.

The basic requirements for the sensory systems can be divided into the following categories:

- Hardware requirements:

  - Design compliant with legislative and operational requirements
  - Resistance to the effects of the industrial environment
  - Can be built into industrial switchboards, as well as a separate realization
  - Compatibility with communication buses
  - Sustainability throughout the life cycle of the control system

J. Ďuďák, G. Gašpar, *Design and Implementation of Sensory Solutions for Industrial Environment*, Signals and Communication Technology,
https://doi.org/10.1007/978-3-031-30152-0_2

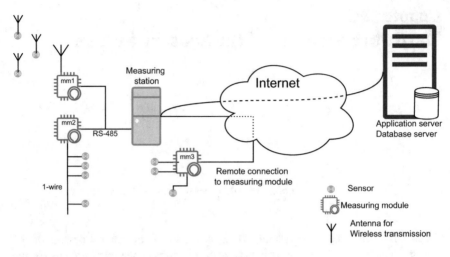

Fig. 2.1 Physical architecture of the sensory system

- Software requirements:

    - Firmware and API designed following standards
    - Compatibility with communication protocols
    - Built-in functionality for configuration and diagnostics
    - API is available for integration into various control systems

- Functional requirements:

    - Device dimensions
    - Supply voltage range
    - Device coverage level
    - Installation regulations

The principal architecture of the sensory system is shown in Fig. 2.1. The measuring module is specially designed hardware that communicates with sensors/sensor modules. The measuring station is a local control computer that controls the measuring module. The measured data are transferred to the data storage.

The system consists of a central server (application server and database server) on which the measured data from a particular installation is stored. The second part of the system is the measuring station. In the system, several measuring stations can be independent of each other, and the data are stored in separate databases. Measuring modules are connected to the measuring station. According to the specification described in [1], the measuring module represents a communication node of the MultiSlave type, so it includes up to 15 submodules. Dozens of measuring modules can be connected to one station. The measuring modules are connected to the measuring station either by the RS-485 serial bus (*mm1* and *mm2* modules) or utilizing a remote connection via a converter module that maps the network IP address to the serial interface. Individual sensors can be connected to the measuring station in one of three ways:

1. Part of the measurement module is one of the 15 submodules (submodules of the *mm3* measurement module in Fig. 2.1).
2. They are connected to a common bus. If it is necessary to connect more than 15 sensors to the measuring module, or the distribution of the sensors is on a larger area, it is advantageous to use a 1-wire bus, which requires only two wires (data and zero wire) for communication and power supply of the connected sensors—measuring module *mm2*.
3. Connection using wireless technology. Certain types of sensors can be used in autonomous sensor mode, which consists of the end sensor itself, power management, and a radio transmitter module—*mm1* measuring module. In this case, the measuring module must contain a submodule for receiving and decoding the radio frequency data packets.

The development of the hardware components described in this book can be divided into two subgroups: development of measurement modules and development of special sensory modules.

The measurement and sensor module solutions were implemented on the ARM CORTEX-M hardware platform. Specifically, 32-bit STM32F0x/STM32L0x microcontrollers were used. Those microcontrollers have a sufficient number of peripherals (UART, I2C, SPI, ADC, DAC, USB) and a well-scalable performance. After designing the connection of a specific module, a service firmware was programmed for each module, which ensured the measurement itself and communication with the necessary peripherals. The firmware is programmed in the ANSI C language, and the GCC compiler was used. For communication with the measuring station, the uBUS [2] communication protocol was designed and implemented, which is derived from the MODBUS protocol and whose model and verification are described in [1, 3, 4]. The protocol works on the master/slave communication principle. Each measurement module provides a set of standard and specific functions. The standard functions are the same for all the measurement modules and include information about the measurement module and the number and types of sensors. Specific functions vary according to the measuring module. They include controlling and reading values from special peripherals, such as a 1-wire bus or a radio frequency communication module.

The proposed sensor system has the following properties:

- The basic architecture is client/server (with the extensions as mentioned earlier).
- The RS-485 industry standard is used as the physical communication layer.
- Communication nodes in this system form measurement modules that contain end sensors or form an interface for connecting further bus expansion. Expansion modules:
  - Module for connecting other communication buses used in industry (e.g., 1-wire)
  - Module for receiving radio frequency data

- Implementation of secure communication at the level of measurement modules.
- Functional hardware solution will be a part of the distributed information system.

  - Independent solution of the measuring stations, data storage, and program application interfaces (API) for data access

- Remote diagnostics of the system hardware components.
- The possibility of integrating the sensor system into IoT-type solutions.

Measuring modules in Fig. 2.1 shown as blocks *mm1*, *mm2*, and *mm3* represent the hardware solutions that have the task of reading the measured values and communicating with a superior measuring station. The standard in industrial communication buses was chosen as the communication standard between the measuring module and the measuring station: RS-485. The RS-485 is a serial communication standard and is primarily used in industrial environments. It is designed to allow the creation of the two-wire, half-duplex, multipoint serial connection. The maximum transmission speed is inversely proportional to the length of the line. The transmission speed for the short connections (up to 10 m) can be up to 10 Mb/s. When communicating over longer distances, the line must be, on both sides, terminated by the termination resistors, i.e., terminators. The terminator should ideally have a value of 120 Ω. Development of measurement modules that are described in this book:

- **OWB** (One-Wire Booster)—module for collecting data from 1-wire bus sensors. It contains an improved dynamic pull-up controller, a 1-wire bus length detector, and an automatic setting of 1-wire bus communication parameters, depending on the bus length.
- **THM** (Temperature and Humidity Module)—module for measurement on four DHTxx-type sensors.
- **RRM** (Radio Receiver Module) receiving module for short-range radio communication. It works with the RTM sensor module.
- **TNZ**[1] (Strain gauge module aka Tensometric module) measurement of bridge connections with application to strain gauge measurements.

2.1   One-Wire Booster Module

The OWB module (Fig. 2.3) represents a 1-wire bus controller with improved features. In Fig. 2.1 this module represents the *mm2* block. The *mm2* measuring module contains a 1-wire bus interface where dozens of sensors can be connected. The block diagram of the module is shown in Fig. 2.2. Experimental measurements have shown that, with a standard connection with a pull-up resistor, the length of

---

[1] The abbreviation is derived from the Slovak name for strain gauge—"Tenzometer".

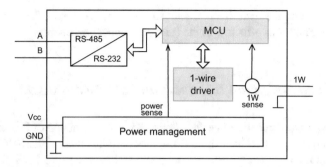

Fig. 2.2 OWB module block scheme

Fig. 2.3 The measuring
OWB module

the 1-wire bus is approximately 100 m with a maximum of 8 connected DS18B20 sensors. The hardware design of the 1-wire driver overcomes these limitations, and experimental measurements have shown that a bus length of more than 500 m with 128 connected DS18B20 sensors can be achieved with reliable communication.

The uniqueness of our solution mainly consists in:

- Using a specialized controller (also a booster) of the 1-wire bus, with which it is possible to use a bus length of up to 500 m
- Detecting the length of the 1-wire bus and automatically setting the bus timing parameters
- Short-circuit detection on the 1-wire bus
- Supports automatic bus timing switching concerning its length
- Communicates with the measuring center via RS-485
- The supply voltage is from 5.5 V to 12 V (version A) or from 7.5 V to 24 V (version B)

The 1-wire technology supports the two ways of connecting the sensors to the bus. The standard wiring uses three wires, where one wire is intended for communication, and the remaining two wires are intended for powering the device. The parasitic mode uses two wires, one wire for communication and the device power supply. The other wire is intended for the GND connection. The OWB (Fig. 2.3) allows using both methods but primarily focuses on the parasitic wiring of sensors. Based on the experimental comparative measurements, this method saves cabling costs and is as effective as the standard connection method.

2.1.1   Communication Interfaces and Microcontroller

The primary communication interface is RS485. For conversion of this interface to RS232 interfaces, with which the used microcontroller works, the SN65HVD10[2] circuit (manufacturer Texas Instruments) is used, shown in Fig. 2.2 RS-485/RS-232 block). This circuit is a half-duplex converter between RS-485 and RS-232. The basic parameters of this circuit, which determine the communication interface limits of the OWB module, are:

- Number of inputs/outputs: 1/1
- Communication mode: half-duplex
- Supply voltage: 3.3 V
- Maximum transmission speed: 32 Mbps
- Overvoltage protection: from -9 V to 14 V
- Maximum working current Icc: 15.5 mA

The OWB module uses the STM32F070F6 microcontroller in TSSOP20 design, i.e., in the case with 20 pins. The basic features of this chip are the following:

- ARM® 32-bit Cortex®-M0 CPU, frequency up to 48 MHz
- Memories: 32 kB FLASH, 6 kB RAM
- 15 I/O pins with 5 V tolerant capability
- One 12-bit, $1.0\,\mu s$ ADC (up to 16 channels)
- 5 timers
- One I2C interface
- Two USART interfaces
- One SPI interface

The use of the existing peripherals of the applied microcontroller is shown in Fig. 2.4.

- Status signaling: LED1—signals when the device is active, i.e., it performs either the measurement itself on the sensors or communicates with the superior device.
- UART—communication interface of the OWB module. It is used in asynchronous mode (USART2_TX/RX).
- For the 1-wire bus driver are used the I/O pins—OW_DATA, OW_UP/ST/DN. This section will be described in detail below.
- Debug interface (SYS_SWCLK/SWDIO)—used only during the solution development.

---

[2] https://www.ti.com/lit/gpn/sn65hvd10.

Fig. 2.4 Utilization of STN32F070 microcontroller pins in the OWB module

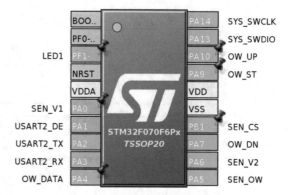

2.1.2   Modified Controller of the 1-wire Bus

The 1-wire bus works with a voltage of 5 V. This voltage represents the inactive state of the bus. With the two-wire connection, the DQ line (communication pin) is used to power the connected sensors, and at the same time, for communication, the VCC and GND pins are connected to zero potential. During the communication, the sensors work thanks to the parasitic power supply—this means that they can only communicate for a certain amount of time. After the end of the communication, the state is restored when the voltage level on the bus corresponds to the value of logical 1. At that time, a circuit that recharges the supercapacitors, which serve as a power source for the sensor, is activated. When the length of the 1-wire bus is greater than one meter, the voltage drop on the bus is observable. In turn, for lengths of more than tens of meters, the capacity of the communication bus has a negative effect. To avoid signal attenuation, in the case when there is no communication on the bus or when there is a need to quickly change the bus level from a low level to a 5 V level, a "hard" 5 V is connected to the bus in order to change its status to the inactive state quickly—that is, to the state logical 1. By using this 1-wire bus controller and adjusting the bus timing (Fig. 1.26), it is possible to achieve a bus length of up to 500 m. Of course, another factor that affects the maximum length of the bus is the number of connected sensors and the bus's own parasitic capacitance.

Adaptive 1-wire Bus Setting   The timing of the 1-wire communication is shown in Fig. 1.26. For the implementation of the 1-wire protocol in the master mode, a modified timing scheme was proposed—Fig. 2.5.

All these time intervals are software configurable. By modifying them, communication on the long 1-wire bus can be adapted. When using a long bus, it is vital to extend the READ_PULSE time to such a value that all the connected 1-wire slave devices can detect the start of communication. Another critical parameter is WRITE_ZERO, to which the master node sends the value "0". Parameters that are set for the standard and long 1-wire buses are given in Table 2.1. The parameters for a long bus were empirically determined for a bus of 600 m in length.

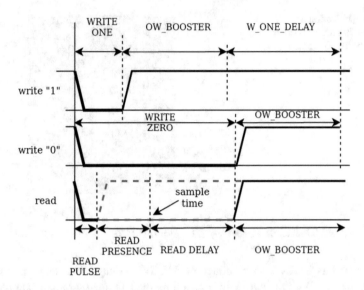

Fig. 2.5 Time parameters of 1-wire communication, master implementation

Table 2.1 Time parameters for the 1-wire bus at standard and long bus lengths

Parameter	Timing for standard length [μs]	Timing for Long bus [μs]
TIME_SLOT	70	85
WRITE_ZERO	45	60
WRITE_ONE	10	12
WRITE_ONE_DELAY	45	60
WRITE_DELAY	10	5
READ_PULSE	6	12
READ_PRESENCE	9	8
READ_DELAY	30	40
OW_BOOSTER	25	25

With a long bus, the OW_BOOSTER time, in which the dynamic pull-up driver is activated, is important. By activating this pull-up driver, we achieve a quick change in the voltage level. In the current solution of the OWB module, the constant value OW_BOOSTER=25 ms was chosen. The other times are derived from this value according to the following relations:

$$\text{WRITE\_ZERO} = \text{TIME\_SLOT} - \text{OW\_BOOSTER} \tag{2.1}$$

$$\text{WRITE\_ONE\_DELAY} = \text{TIME\_SLOT} - (\text{WRITE\_ONE} + \text{OW\_BOOSTER}) \tag{2.2}$$

For the timing of the read operation, the following relationships are applied:

$$\text{READ\_PRESENCE} + \text{READ\_PULSE} = 18 \tag{2.3}$$

$$READ\_DELAY = TIME\_SLOT - (READ\_PULSE + READ\_PRESENCE + OW\_BOOSTER)$$
$$(2.4)$$

WRITE_ONE and WRITE_ZERO times are $10\,\mu s$ and $45\,\mu s$, respectively, when using a short bus (up to 50 m). When using a more extended bus, the signal attenuation and its reflection from the end of the bus appear during communication—this is shown in Fig. 2.7. This reflection causes the level change, from 5 V to 0 V or vice versa, to be delayed for up to 10–$20\,\mu s$. This phenomenon cannot be prevented, but it is possible to adjust the timing of the 1-wire bus (by increasing the value of WRITE_ZERO) so that when reading a bit, there is enough time to lower the bus voltage level reliably. The setting of the 1-wire bus parameters is done using the command CMD_SLAVE_SET_PARAM of the uBUS protocol, which was described in the section "Specification of the uBUS protocol" (Sect. 1.3.3).

2.1.3   Detection of a Short-Circuit on the 1-wire Bus and Current Consumption Monitoring

Short-circuit detection on the 1-wire bus is a functionality of the OWB module to indicate faulty sensors wiring or the accidental connection of the data and ground wires. A short circuit is indicated by a warning LED. When the short circuit is detected, the OWB module communicates. However, the error code DEVICE_HW_ERROR_2 (0x08) is returned for functions connected to the 1-wire bus. The short-circuit detection is realized by connecting a voltage divider, with a resistance ratio of 1:3, directly to the 1-wire bus. This circuit should be designed to minimally affect the communication on the 1-wire bus itself. The resulting resistance of the divider is 133 kΩ. If the active level of the 1-wire bus is 5V, then the current flowing through the voltage divider will be $37\,\mu A$.

Monitoring the current load of the 1-wire bus is essential when using sensors with higher consumption requirements in parasitic mode. Typically, these are units of mA per module. Such a module is described in the section Sensory modules—OWS module, Sect. 2.5.1.

A ZXCT1022 circuit was used to monitor the current load. The ZXCT1022 is a Low Offset High-side Current Monitor used to eliminate the need to disrupt the ground plane when sensing a load current. It provides a fixed gain of 100 for applications where a minimal sense voltage is required. The very low offset voltage enables a typical accuracy of 3% for sense voltages of only 10 mV, giving better tolerances for little sense resistors (R_{sense}) necessary at higher currents. The wide input voltage range of 20 V down to as low as 2.5 V makes it suitable for various applications. A minimum operating current of just $25\,\mu A$ [5].

In Fig. 2.6 is shown a wiring diagram for current consumption monitoring.

The value for the R_{sense} resistor was chosen to be 50 mΩ. There is a relationship between the output voltage and the actual current:

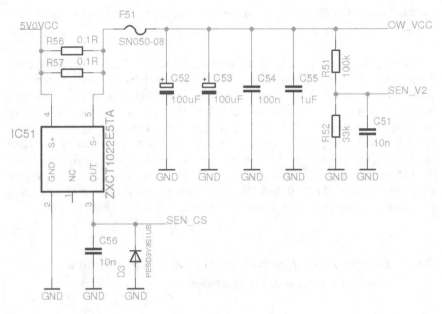

Fig. 2.6 Monitoring the current load of the 1-wire bus

$$V_{out} = I * R_{sense} * k \tag{2.5}$$

where $R_{sense} = 0.05\,\Omega$ and $k = 100$ (amplification of the used ZXCT1022 circuit). For the flowing current, the following relation holds:

$$I = \frac{V_{out}}{k * R_{sense}} \tag{2.6}$$

After the calculation: $I = V_{out}/(100 * 0.05\Omega) = V_{out}/5\Omega$.

The V_{out} value is read off by the ADC converter in the used microcontroller. The pin SEN_CS used for this functionality is shown in Fig. 2.4.

2.1.4   The Bus Length Detection

Physical processes influence communication when using a 1-wire bus longer than 200 m. This is a reflection of the signal from the end of the bus. The signal propagation speed in the conductor depends on several factors, like the used conductor material or the bus impedance. This speed can be from 0.5 to 0.9 of the speed of light. We consider the signal propagation speed in the conductor to be (3/5) of the speed of light. We found this value empirically for the used driver. The speed of signal propagation in the conductor used has a value:

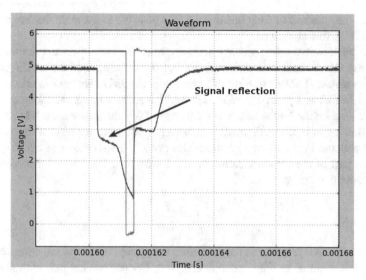

Fig. 2.7 Signal reflection on a long bus

$$V = \frac{3}{5} * 0.299 \frac{m}{ns} = 0.18 \frac{m}{ns} \tag{2.7}$$

For a 180 m long bus, this represents 1 ms per bus length. We will observe the reflection of the signal from the end of the bus as a delay of 2 ms, since the signal must pass through the bus twice. The reflected signal returns to the beginning of the bus in 2 ms. When monitoring the course of the signal, this is manifested as a specific retention of the signal (Fig. 2.7). A 1-wire bus with a length of 500 m was used for testing. From relation 2.7, the delay for a bus length of 500 m will be 2.7 ms. The waveform measured at the beginning of the bus is shown in Fig. 2.7. The total length that the signal must overcome is 2*500 m = 1000 m, representing a signal reflection delay of 5.4 µs, corresponding to the measured value. The green line in Fig. 2.7 is the value measured from the level comparator that is applied to the 1-wire bus signal. For the correct interpretation of the logical levels, it is necessary to set the comparative levels appropriately.

The 1-wire timer algorithm was designed for the automatic adaptive setting of appropriate timing parameters of the 1-wire bus. This algorithm was implemented in the firmware of the OWB measuring module. The principle of the algorithm is as follows: at the start of the module, several cycles of changing the voltage level of the 1-wire bus are performed, while the shape of the voltage response on the bus is measured using a fast A/D converter. The time it takes for the voltage level of the 1-wire bus to reach the value, after which the voltage starts to decrease more rapidly, to the value of 0 V, is proportional to the bus length (the parameter is the type of bus conductors). Based on this finding, the measurement module automatically

chooses the standard or long bus timing. It is possible to estimate the length of the
bus utilizing this algorithm.

The operation for determining the length of the bus is started automatically
when the OWB module is switched on, but the command of the uBUS protocol
can also invoke it. After starting this operation, the A/D converter is activated—in
Fig. 2.2, which is a part of "1W sense"—which starts a quick measurement of the
voltage level on the 1-wire bus. The frequency of this measurement is 4 MHz, so the
resolution is $t = 0.25$ ms. Figure 2.8 shows a graphic representation of the measured
course with the 1-wire timer algorithm. This process is also shown in Fig. 2.7, which
shows a part of the communication on the 1-wire bus. The basic formula used for
calculating the bus length is:

$$l = \frac{vt}{2} \tag{2.8}$$

where l is the bus length, t is the measured time, and v is the propagation speed of
the electrical signal in the conductor material. Copper twisted pair with shielding
was used as material. After several experiments with the cabling with a known buss
length, the value is expressed as:

$$v = 0.6c \tag{2.9}$$

where c is the speed of light. The waveforms in Fig. 2.8 represent the buses of
lengths 202 m and 404 m, respectively. The times shown have values: $t_1 = 2.25$ ms
and $t_2 = 4.5$ ms.

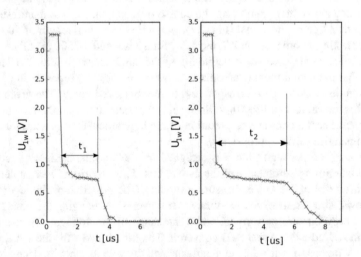

Fig. 2.8 Measured delay of the 1-wire bus level change

2.1.5   *Improved Active 1-wire Bus Pull-up*

Figure 1.36 shows the solution for implementing an active pull-up driver for the 1-wire bus. The OWB hardware module contains the improved version of this design. The wiring diagram of this design is shown in Fig. 2.9.

The main differences between the OWB module solution and the standard controller with an active pull-up driver are the following:

- Selection of parts with better parameters for the given purpose:

  – The used transistors are designed for switching and controlled from the logic circuits. The BSS84 and 2N7002 transistors are used in the original design, while the IRLML2244 ones are used in the OWV design. The used transistors have smaller switching losses.
  – The transistor's transition resistance (S-D) in the switched state is significantly lower—54 mΩ for our modified design vs. 10 Ω—for the original design.

- Special integrated circuits (U61) with a charge pump are used to control the switching transistors for a faster change of the state of the transistor. This part is absent in the original design.

  – U61—The MCP14E11 device is the high-speed MOSFET driver, capable of providing 3.0 A of the peak current. The dual inverting, dual non-inverting, and complementary outputs are directly controlled from TTL or CMOS (3–18 V). These devices also feature low shoot-through current, near-matched rise/fall times, and propagation delays, which make them ideal for high switching frequency applications.

- Protection of connecting circuits against unwanted overvoltage on the bus implemented by a special diode (D53) combined with a fast diode (D52).

  – D53—This diode is explicitly designed to protect sensitive electronic equipment from voltage transients induced by lightning and other transient voltage events. Fast response time: typically, less than 1.0 ps from 0 Volts to VBR.
  – D52—The MBR0530 uses the Schottky Barrier principle with a large area metal-to-silicon power diode. They are ideally suited for low voltage, high-frequency rectification or as free-wheeling and polarity protection diodes in surface mount applications, where compact size and weight are critical to the system.

- The detected signal (OW_O) is shaped by a fast comparator—U71.

  – U71—the MCP14E11 circuit is used as a comparator.

- For the sampled signal from the bus (OW_O), a window comparator (IC81A) is connected with a fast operational amplifier for shaping this signal.

  – IC81A—This comparator has extremely high input impedance (typically greater than $10^{12}\Omega$), allowing direct interfacing with the high-impedance

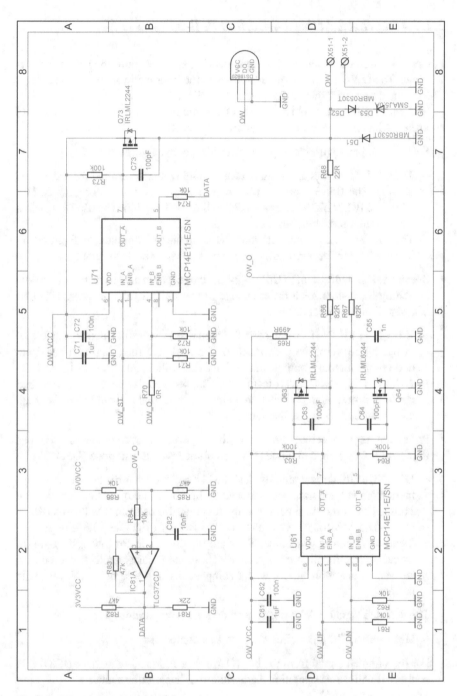

Fig. 2.9 Enhanced 1-wire driver schematics

sources. The outputs are n-channel open-drain configurations and can be connected to achieve positive-logic wired-AND relationships.

- The supply voltage for the "OW" bus is isolated by a polymer reverse fuse (F51, Fig. 2.6) to protect the supply of control circuits.
- The wiring is supplemented by a battery of capacitors with a low "ESR" (C5x, Fig. 2.6) to cover the short-term impulse withdrawals.
- Scanning the current taken by the "OW" bus—SEN_CS (Fig. 2.6)—to evaluate the error condition when the current flows through the bus beyond the permitted limit. Such a situation can occur when using several One-Wire-Slave modules.
- Voltage scanning on the "OW" bus to detect a short circuit or low voltage on the 1-wire bus. Signal SEN_V2 (Fig. 2.6). Such a condition can occur in an accidental short circuit or when a damaged sensor is connected.

2.1.6 Mechanical Implementation of the 1-wire Bus Sensors

The recommended wiring of sensors on the bus is linear (Fig. 1.25), where all the sensors are connected to one bus with a maximum branch length of 3 m. Before the physical installation of sensors in the industrial environment, there were different applications designed for several applications. In the described applications, the DS18B20 sensors were used. DS18B20 is a digital thermometer whose physical version is shown in Fig. 2.10. The basic parameters of this sensor are:

- Can be powered from data line; Power supply range is 3.0–5.5 V
- Measures temperatures from $-55\,°C$ to $+125\,°C$
- $\pm0.5\,°C$ accuracy from $-10\,°C$ to $+85\,°C$
- Converts temperature 12-Bit word in 750 ms (max)
- User-Definable nonvolatile (NV) alarm settings

2.1.7 Implementation of Sensors for Ambient Temperature Monitoring

The first draft of the mechanical design consisted of a plastic case in which the sensor itself was stored. The sensor housed in this case was electrically isolated and hermetically sealed in the body of the case. The disadvantage of this solution was the case material itself because it formed thermal insulation for the sensor. This plastic case was replaced by a brass case in the subsequent development iteration (Fig. 2.11).

The brass cases were treated with nickel plating for further resistance increase, especially to chemical and environmental stress. The manufacturing process of the sensor remained similar to that of the sensor in the plastic case.

Fig. 2.10 Physical properties
of the DS18B20 (**a**) TO-92,
(**b**) SO, (**c**) μSOP sensors [6]

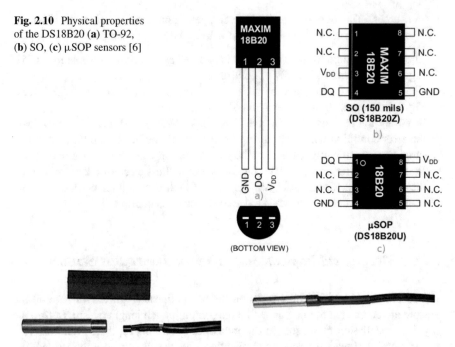

Fig. 2.11 Temperature sensor with metal cover

2.1.8   Unique Designs for 1-wire Temperature Sensors

The temperature sensor was built into the building structures. This is a proposed solution for storing a temperature sensor on an asphalt road. A unique mechanical design had to be created to prevent the destruction of the sensor during its installation. Before designing the mechanical version, it was necessary to verify whether the used sensor would not change its parameters under the increased temperature during the asphalt laying, which reaches up to 190 °C. The use of such sensors in the diagnostics of traffic structures is described in [7], especially regarding using the measured data as input data for the prediction model of the icing formation on asphalt roads. The sensor cover was designed, as shown in Fig. 2.12, and its 3D model was created, which was then subjected to a simulation of the mechanical stress caused by passing cars. After incorporating the simulation results, an experimental series of sensors was created from the 3D model built into the road. The design of such a durable sensor proved correct and suitable for use in the demanding conditions of the construction industry, as it has been working reliably in several installations since 2016. Photo documentation of the designed sensor installation on the road is shown in Fig. 2.13. The sensor body was subsequently covered with asphalt.

Fig. 2.12 Section of the
temperature sensor cover

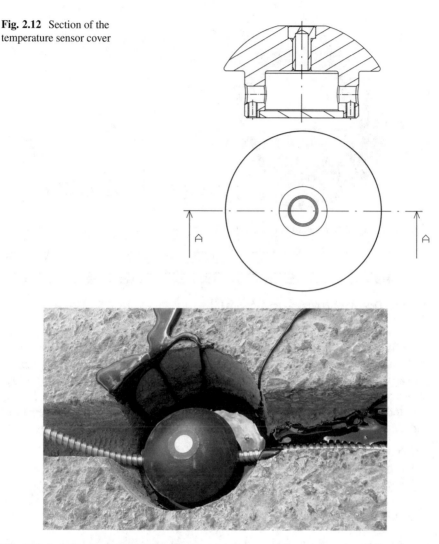

Fig. 2.13 Installation of a durable temperature sensor

Implementation for Measuring the Vertical Temperature Profile A vertical
temperature sensor, shown in Fig. 2.14, was designed for other temperature measurement applications. Similar to the previous case, the design uses a DS18B20
digital thermometer. It is a sensor for the applied measurement of the vertical
temperature profile, with application mainly in the construction area. The original
design was predicted using commonly available components, from which the
verification series was created. A polycarbonate strip was used as the supporting
profile of the individual sensors. Sensors were attached at a strip in defined distances
of 100 mm. The sensors were soldered into a series-parallel connection in the

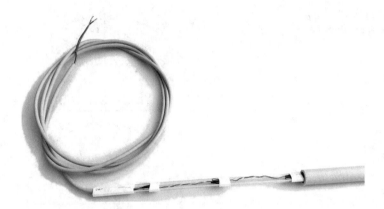

Fig. 2.14 Vertical temperature sensor design [8]

Fig. 2.15 Design of the printed circuit boards [8]

parasitic mode. The cover of the vertical sensor was made of a polypropylene tube of 20 mm in diameter and a wall thickness of 2.7 mm. After inserting the support profile with the sensors and the connected cabling into the housing, the vertical sensor was hermetically sealed with a polyurethane sealing compound. During the production of the verification series, several shortcomings of the implemented vertical sensor were detected. It was mainly about the following deficiencies:

- Laborious production
- Unequal distances between individual sensors given by design
- Dimensions of supporting materials
- The wall thickness of the vertical sensor housing and the heat transfer affected by it

The Dbar Probe Modifications of the construction and mechanical design of the sensor were proposed considering the shortcomings mentioned above. The first three shortcomings were eliminated by the design of printed circuit boards, shown in Fig. 2.15. These are the modular boards with dimensions of 50, 100, and 250 mm, from which the necessary length of the vertical sensor can be assembled by soldering at the connection points DQ and GND.

The distance between the individual plates, intended for the connection of the DS18B20 sensors, was set at 50 mm, but it can be further reduced by simply modifying the design of the plates. The last shortcoming, represented by the wall thickness of the cover, was eliminated by using a different cover with an outer diameter of 10 mm and a wall thickness of 1 mm. Based on the design of the necessary installation elements—plug, casting funnel, and funnel's cover, their 3D

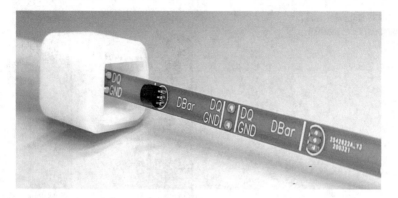

Fig. 2.16 Implementation of the DBar vertical temperature probe design [8]

model was created. A 3D model was subsequently printed on a 3D printer. The functionality of the individual elements was verified when the vertical sensor was sealed with a polyurethane sealing compound. Figure 2.16 shows how to insert the prepared printed circuit boards with soldered thermometers into the case.

When designing the DBar probe, emphasis was placed on the appropriate material selection for the case itself because this material can affect heat transfer from the external environment to the sensor itself. A detailed analysis was published in the article [8].

Two experimental probes[3] were proposed for numerical simulation of the heat transfer from the environment and experimental verification of obtained results. Both probes are 250 mm long, and the sensors are placed on the printed circuit boards at a 50 mm distance. The experimental probes, shown in Fig. 2.17, differ in the applied housing material, dimensions, and the number of sensors used for the proposed experiments. The decision regarding the housing material was made based on its availability on the market, considering mainly its environmental resistance. Considering this fact, the PPR and HDPE plastic materials were selected. These plastic materials offer high abrasion and corrosion resistance to soil chemicals. Both materials are also suitable for operating temperature, as the experimental setup is supposed to be installed in the forest, not in a laboratory environment. Another reason is the appropriate thickness of the available plastic tubes. This property affects the heat transfer from the environment to probes attached inside the housing. In the case of probe 1 (PPR probe in Fig. 2.17a), the probe housing is a polypropylene plastic tube of a 20 mm diameter and a wall thickness of 3.2 mm. Sensors marked 100, 110, and 120 are placed in the body of the probe with an offset of 4.5 mm from the printed circuit board, with a location corresponding to the thermometers 00, 10, 20. Sensors marked 01, 02, and 03 are placed on the

---

[3] This section describes the authors' solution, which was in the past partially published in Applied sciences journal—[8].

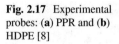

Fig. 2.17 Experimental probes: (**a**) PPR and (**b**) HDPE [8]

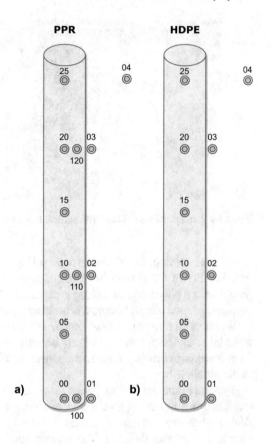

body of the probe with the location corresponding to the thermometers 00, 10, and 20, respectively. The sensor marked 04 is, as in the case of probe 2, an auxiliary thermometer for measuring the ambient temperature. The base construction of probe 2 (HDPE probe in Fig. 2.17b) uses thermoplastic polymer (HDPE material) as a tube housing material of a 10 mm diameter and a wall thickness of 1 mm. Sensors DS18B20 marked 00, 05, 10, 15, 20, and 25 are in the probe's body with a polyurethane sealant. Sensors marked 01, 02, and 03 are placed on the body surface of the probe with the location corresponding to the thermometers 00, 10, and 20. The sensor marked 04 is an auxiliary sensor for measuring the ambient temperature.

There are sensors inside the tube, as we have already mentioned. These probes are encapsulated using the two components consisting of a resin and curing agent. The PUR sealant is used to isolate electric components from the surrounding environment to prevent damage caused by water or air humidity. Equally important, the dissipation factor using the PUR sealant is acceptable. The proposed experiments covered two basic types of measurements:

(A) Measurement in a water bath to verify the measurement procedure during the step temperature change.

1. The measurement starts at ambient temperature.
2. The probe is immersed in a water bath at approximately 55 °C (domestic hot water).
3. The measurement is completed after approaching the temperatures measured by sensors 00, 05, 10, 15, 20, and 25 and the auxiliary thermometer 04.

The measurement is repeated at different values of the water bath temperature (30 °C and 10 °C). The time resolution of the measurement is assumed to be ∼1 s.

(B) Measurement in a climate chamber to verify the measurement procedure in the case of gradual temperature change.

1. The probe is placed into the climate chamber, and the measurement starts.
2. The temperature increases by 10 °C considering that the initial temperature is adjusted in the climate chamber, followed by a dwell time of approximately 6–8 min at maximum temperature. Then, the climate chamber is switched off.
3. The measurement is completed after approaching the temperatures measured by sensors 00, 05, 10, 15, 20, and 25 and auxiliary thermometer 04.

The measurement is repeated at different maximum temperatures of the climate chamber (plus 15 °C and 25 °C). The time resolution of the measurement is assumed to be ∼ s.

Simulation Results   Figures 2.18 and 2.19 illustrate the temperature distribution in probes 1 and 2 at 30 s, 60 s, 90 s, and 300 s after immersing the probe in a water bath at approximately 55 °C. The temperature differences in probe 2 are more minor than in probe 1. At a time of 300 s, the temperatures in probe 2 vary from 52 °C to 56 °C. Probe 1 has temperatures from 48 °C to 56 °C, while the temperatures of the PUR sealant are from 48 °C to 52 °C. The response of probe 1 to the sudden temperature change is more delayed.

Due to its construction from robust material and the volume of sealing material used, Probe 1 registers changes in the ambient temperature more slowly and is, therefore, suitable for installations in environments where slow changes occur. Its advantage is the higher mechanical resistance the selected semi-finished product provides for case manufacturing. Probe 2 has been shown to register changes in ambient temperature more quickly and is, therefore, suitable for installations in environments where such rapid changes occur. Its disadvantage is the lower mechanical resistance caused by the used case material.

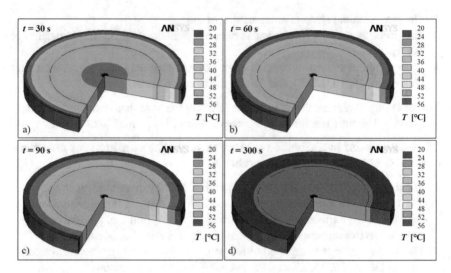

Fig. 2.18 Temperature distribution in the probe 1 at time of (**a**) 30 s, (**b**) 60 s, (**c**) 90 s and (**d**) 300 s after immersing the probe in a water bath at approximately 55 °C [8]

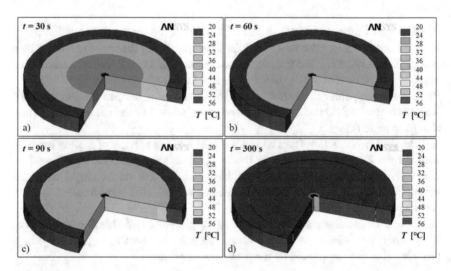

Fig. 2.19 Temperature distribution in probe 2 at times of (**a**) 30 s, (**b**) 60 s, (**c**) 90 s, and (**d**) 300 s after immersing the probe in a water bath at approximately 55 °C [8]

2.2 Measuring Module RRM[4]

The primary motivation for developing such a measuring module was designing a wireless communication system to monitor the environmental parameters of the bulk materials, namely wood chips. It is impossible to use sensors connected to a fixed bus line in this situation. The architecture of the wireless RF monitoring system is shown in Fig. 2.20 [10].

In the proposed monitoring system, n environmental sensors are located in the monitored area. As the monitored area may be larger than the range of RF transmitting modules, the system contains m receiving modules. Each receiving module can receive data from a set of sensors, while these sets may overlap. For the implementation of wireless communication, technology was chosen that allows communication at a distance of a minimum of 50 m and automatic reception of data from transmitter probes without requiring further configuration. WI-FI or Bluetooth communication would require connecting the transmitters to the wireless network. The use of RF signal in the free RF band 433.92 MHz appears to be optimal for use in terms of implementation and energy-saving. When analyzing the problem, the requirements for the entire monitoring system were defined as follows:

- Temperature measurement in the range $-10\,°C$ to $90\,°C$ with an accuracy of $\pm0.5\,°C$ and resolution of $0.1\,°C$
- Autonomous operation of measuring probes for at least three months
- Probe design resistant to the pressure of stored material in which the probe is placed and against the mechanical stress caused by mechanisms used in the processing and material handling
- Operation of probes in the free RF band 433.92 MHz
- One-way communication protocol, which allows the detection of a duplicate multi-probe transmission on the receiver side

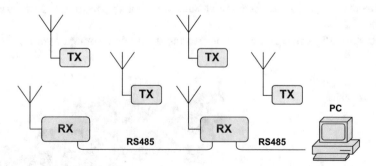

Fig. 2.20 Architecture of the RF monitoring system [9]

---

[4] This section describes the authors' solution, which was in the past partially published in Journal of Sensors—[9].

- Parameterized probe transmission frequency depending on the temperature of the material
- Reception of a signal covering the entire storage site, secured by multiple receivers with duplicate measured data filtering
- The data model of the system—the internal model of the measurement system maintaining information on the hardware configuration, active sensors, their physical location, measured data, and other supplementary information
- Secure data transmission from probes to receiving modules, secure communication from the receiving modules and the control computer, secure communication between the client software and the control computer
- Platform independent, simple client software
- Remote access to data

The RRM module (Radio Receiving Module)—Fig. 2.21—represents a receiving station for autonomous RTM RF sensors (described in the Sect. 2.5—Sensory modules). The connection design consists of two essential parts—a standard communication, a power supply, and a computing part with a microcontroller and a radio signal receiver module (Fig. 2.22). The advantage of such a solution is the possibility of replacing the receiver with a different frequency. The frequency 433.92 MHz is used in this connection and is freely available for general use if certain conditions are met. The main features of the module are:

- Supports different received frequencies through the module replacement
- Reception and storage of values from 128 different probes
- Each probe can contain 5 different measured quantities
- Works with a radio data transfer rate of 1000 baud
- Communicates with the measuring center via RS-485
- The supply voltage is from 7.5 VDC to 28 VDC

Configuration of the module according to the multiSlave specification is:

- sub-slave 0—radio interface that ensures signal reception from the RF modules (Sect. 2.5.2).
- Sub-slave 15—internal memory for storing the module configuration.

Fig. 2.21 The RRM measuring module

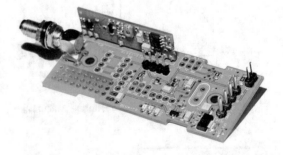

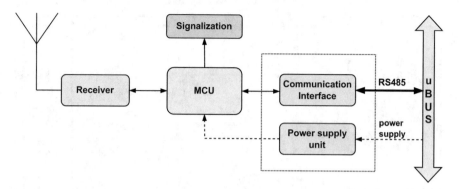

Fig. 2.22 Principal block diagram of the RRM module

The solution has an easily editable firmware part that enables the implementation of various sensors with RF communication. Such a solution is especially advantageous when integrating several types of sensors from different manufacturers. The most common use is in monitoring biologically active materials that require operator intervention when the critical parameters are reached (e.g., flash point).

The circuit diagram of the RRM module is shown in Fig. 2.23. The STM32F030 microcontroller was used in the implementation, and the AC-RX2/CS hardware module from Aurel Wireless[5] was used as the radio module. The properties of this module are listed in Table 2.2.

The receiving module (marked as RF1 in the diagram) has no communication protocol implemented. Its output is a demodulated RF signal, which is subsequently processed in a microcontroller. The output from the RF1 module is the pin labeled DATA. The level of this signal is 5 V; therefore, a simple resistor divider (diagram, part B3) is used, which transforms the voltage level to 3.3 V. The RRM module was designed considering its easy expandability in mind. For this reason, the standard interfaces, which the microcontroller contains, were introduced on the printed circuit board. Those are the two independent I2C interfaces—connector JP31 and JP32 (diagram—A5, B5), SPI—connector JP22 (diagram—B1), UART—connector JP24 (diagram—D1), and STLINK microcontroller programming interface—connector JP23.

2.2.1 The RF Communication Layer

The lowest-level communication protocol RH_ASK was selected from the Radio-Head library.[6] RadioHead is a Packet Radio library for embedded microprocessors.

---

[5] https://www.aurelwireless.com/receivers/.

[6] https://www.airspayce.com/mikem/arduino/RadioHead/.

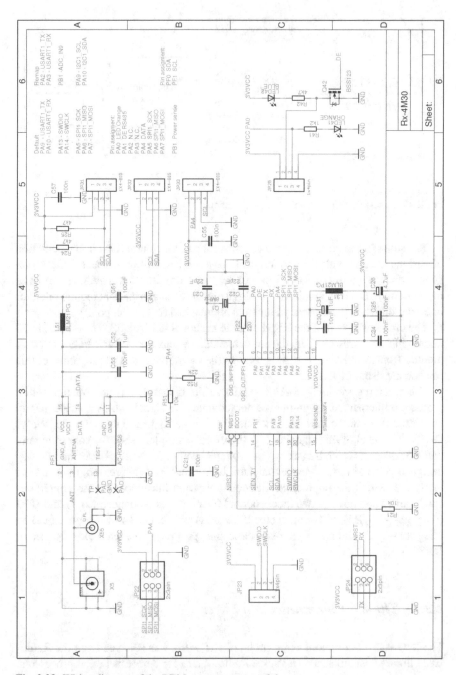

Fig. 2.23 Wiring diagram of the RRM measurement module

Table 2.2 Properties of module Aurel Wireless

Characteristics	Value
Working frequency	433.92 MHz
Modulation	OOK (On-off keying)
Used filter	SAW
Supply voltage	4.75–5.25 V
Supply current	2.75–3 mA
RF sensitivity	−103 to −107 dB
Operating temperature	−20 to +85 °C
EU norm	Compatible with RED 2014/53/EU[a]

[a] https://eur-lex.europa.eu/legal-content/EN/TXT/?uri=CELEX:32014L0053

Table 2.3 The format of the RF data packet

PREAMBLE	START	LED	DATA_VW	FCS
36 bit	12bit	1B	1-27B	2B

This library provides features to send short messages without addressing, retransmitting, or acknowledging, a bit like the UDP over wireless, using the OOK. It supports several inexpensive radio transmitters and receivers. The maximum length of the data part is defined by **VW_MAX_PAYLOAD** (27 bytes). Each byte in the sent packet is encoded using 4-to-6 encoding for the two 6-bit words. The purpose of 4-to-6 encoding is to achieve the state in which the same number of ones and zeroes will be transmitted in the broadcast. The format of the data packet is shown in Table 2.3.

The data packet parts are:

- PREAMBLE—36-bit training preamble consisting of 0-1 bit pairs
- START—12 bit start symbol 0xB38
- LEN—1 byte of message length byte count (4–30), count includes byte count and FCS bytes
- DATA_VW—n message bytes, maximum n is 27 (VW_MAX_PAYLOAD)
- FCS—2 bytes Frame Check Sequence

A description of the communication protocol application for the RF probes is given in the Sect. 2.5.2.

2.3 Measuring Module THB

The measuring THB module (Temperature-Humidity Board), shown in Fig. 2.24, represents the interface for connecting 4 sensors-measuring temperature and relative air humidity. The wiring design uses a standard communication, power supply,

Fig. 2.24 Measuring DHT
board

and computing part with a microcontroller and a part for connecting a maximum of 4 DHT11/DHT22 sensors. Sensors are connected using a 4-channel analog multiplexer 74HC4052.

The essential features of the proposed DHT solution can be summarized as follows:

- Supports connection of 4 AM2302/DHT22 sensors[7]
- Measuring frequency: 0.5 Hz
- Supports the bus length up to 20 m for each connected sensor
- Communication bus: RS-485
- The supply voltage is from 7.5 VDC to 24 VDC

Configuration of the module according to the multiSlave specification:

- sub-slave 0—sensor 1 for relative air humidity
- sub-slave 1—sensor 2 for relative air humidity
- sub-slave 2—sensor 3 for relative air humidity
- sub-slave 3—sensor 4 for relative air humidity
- sub-slave 4—sensor 1 for temperature (additional measurement)
- sub-slave 5—sensor 2 for temperature
- sub-slave 6—sensor 3 for temperature
- sub-slave 7—sensor 4 for temperature
- sub-slave 15—internal memory for saving the module configuration

The circuit diagram of the module is shown in Fig. 2.25. The basis of the measuring module is the microcontroller STM32F030 (diagram—IC21, position C3) and the multiplexer 74HC4052 (diagram IC51, position C5). There are 4 lines (DHT1 to DHT4) for connecting the sensors to the multiplexer IC51. The sensors are switched using the multiplexer A, B address inputs. These inputs are controlled by pins PA2 and PA3 of the used microcontroller. Since the microcontroller and the multiplexer have different logic levels, it is necessary to use voltage level matching in their mutual communication. In the diagram in Fig. 2.25, converters between 3 V and 5 V logic are connected in part A3 and A4 of the schematic. Like the RRM

---

[7] https://datasheetspdf.com/pdf/942482/ETC/AM2302/1.

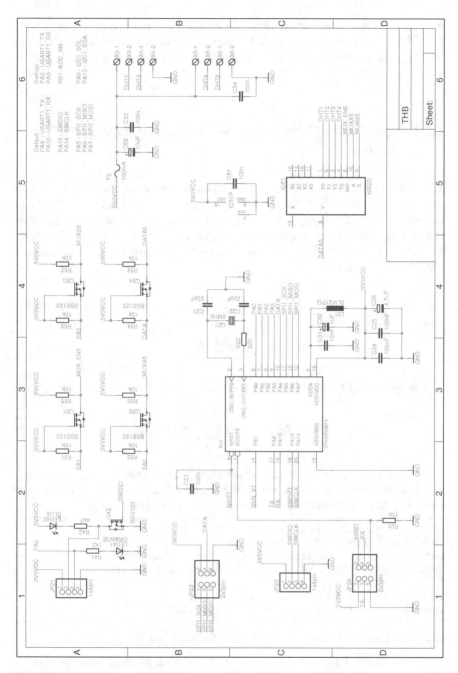

Fig. 2.25 Wiring diagram of the THB measurement module

module, the THB module has the basic communication buses on the printed circuit board for easier module expansion. Sensors are connected via connectors X3 and X5. According to the manufacturer, the maximum bus length of the sensor itself is 100 m.

2.3.1   Application Communication Interface of the AM2303 Sensor

The AM2303 sensor has a digital output and its own protocol, similar to the 1-wire protocol. The output is a 40-bit word consisting of 3 parts: RH data (Relative Humidity)—16-bit, T data (Temperature)—16-bit, and Check Sum (8-bit). The data is coded as integers. For temperature, the highest bit has the meaning of the sign. The following relations (Eqs. 2.10 and 2.11) apply to the calculation of temperature and relative air humidity:

$$T = \frac{data_T}{10} \tag{2.10}$$

$$RH = \frac{data_{RH}}{10} \tag{2.11}$$

The relation 2.12 is used to calculate the checksum:

$$CheckSum = Byte \left(\sum_{n=0}^{3} DATA[i] \right) \tag{2.12}$$

where DATA represents the data part of the response, i.e., the first 32 bits, or 4 Bytes, respectively, *Byte* represents the first 8 bit of the result. In the Listing 2.1, there is a sample of the sensor's response and conversion to the required values:

```
DATA    = 00000010 10001100 10000001 01011111 01101110
DATARH  = 00000010 10001100 => 652
DATAT   = 10000001 01011111 => sign="-", value=351
CHECK   = 00000010+10001100+10000001+01011111 = 101101110
RH      = 652/10 = 65.2
T       = -351 / 10 = -35.1
CheckSum = first 8 bit of CHECK = 01101110
```

Listing 2.1 Example of value computation for the AM2303 sensor

When the MCU sends a start signal, AM2302 changes from standby-status to running-status. When MCU finishes sending the start signal, AM2302 will send a response signal of 40-bit data that reflects the relative humidity and temperature to MCU. The AM2302 will change to standby-status when the data collecting is

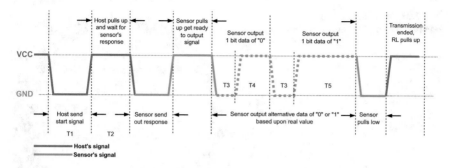

Fig. 2.26   AM2303 sensor communication start

finished and if it does not receive the start signal from the MCU again. In Fig. 2.26 is an example of communication between MCU and AM2303.

Start of communication:

- MCU will pull low data bus, and this process must last at least beyond 1∼10ms (T1).
- MCU will pull up and wait for 20–40 μs for the AM2302's response (T2).

Sending data from the sensor:

- Data is sent bit by bit; each bit starts by lowering the voltage level to logical 0 for at least 50 μs (T3).
- If the value "0" was sent, the bus level changes to logical 1 for 26–28 μs (T4).
- If the value "1" was sent, the bus level will change to logical 1 for 70 μs (T5).

End of communication

- The AM2303 sensor will reduce the bus voltage level to logical 0.
- After the end of the reception, the MCU will raise the level to logical 1.

2.3.2   Implementation of Sensors for the THB Measuring Module

The construction of the temperature and relative air humidity sensor is shown in Fig. 2.27. An AM2302 sensor with a digital output was used as a relative air humidity sensor. The parameters of the used sensor are:

- 3–5 V power and I/O
- 2.5 mA max current use during conversion (while requesting data)
- Calibrated digital signal
- Humidity readings (0–100%) with 2–5% accuracy

Fig. 2.27 Temperature and
relative air humidity sensor

- Temperature readings (-40 to $80\,°C$) with $\pm0.5\,°C$ accuracy, full range temperature compensated
- Long transmission distance: up to $100\,m$
- Sampling rate: $0.5\,Hz$ (once every $2\,s$)
- Body size $27\,mm \times 59\,mm \times 13.5\,mm$ ($1.05'' \times 2.32'' \times 0.53''$)

The final version of the sensor was realized in a duralumin cover with perforation for air flow to maintain the correct functionality of the relative air humidity sensor.

2.4 The Measuring TNZ Module

The measuring module TNZ (Strain Gauge Measuring module aka Tensometric Module), shown in Fig. 2.28, serves as a two-channel converter for the bridge connections. An added feature is the implementation of a simplified 1-wire bus driver for applications up to $150\,m$. The wiring design uses a standard communication, power supply, and computing part with a microcontroller, a 1-wire exciter, and a part for connecting a maximum of two devices in a bridge connection. An example of use is the measurement of a strain gauge in a full bridge. Another example is a PT1000 thermometer in a 4-wire connection. The temperature can be measured using the DS18B20 sensors as an additional measurement.

The essential features of the proposed TNZ solution can be summarized as follows:

- Supports measurements with a frequency of up to $16.6\,Hz$
- Enables the use of 20-bit or 24-bit AD converters
- Allows measuring bridge connections (for example, strain gauges, PT thermometers)
- Communication interface: RS-485

Fig. 2.28 The measuring
TNZ board

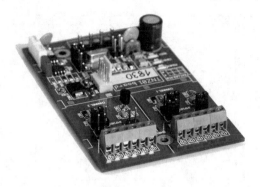

- The supply voltage is from 7.5 VDC to 24 VDC

Configuration of the module according to the multiSlave specification:

- sub-slave 0—converter for the first channel
- sub-slave 1—converter for the second channel
- sub-slave 15—internal memory for storing the module configuration

The TNZ module contains 2 equivalent channels with analog input. The input channel wiring diagram is shown in Fig. 2.29. Each channel contains an AD7780[8] analog-to-digital converter, which has the following features:

- Bit resolution: 24 bit
- AD converter configuration: using external pins:
  - pin GAIN: amplification of the input signal: 1, 128
  - pin BPDSW: power down and reset
  - pin FILTER: define output data rate (10 Hz, 16.6 Hz)
- Output data rate: 10 Hz, 16.6 Hz
- RMS noise:
  - gain = 128: 44 nV (10 Hz); 65 nV (16.6 Hz)
  - gain = 1: 2.4 µV (10 Hz); 2.7 µV (16.6 Hz)
- Power supply: 2.7–5.25 V
- Current:
  - gain = 1; 11 µA
  - gain = 128; 330 µA
- Communication interface: read-only SPI

---

[8] https://www.analog.com/media/en/technical-documentation/data-sheets/ad7780.pdf.

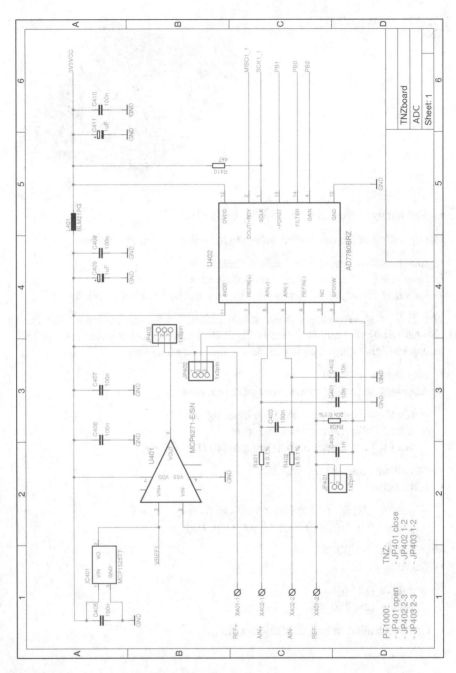

Fig. 2.29 Connecting of the TNZ module input channel

Configuration of the transducer wiring for the strain gauge or temperature sensor use is done by physically changing the wiring using switches on the TNZ module printed circuit board:

1. Measuring on the PT100 sensor

   - JP401: open
   - JP402: 2-3
   - JP403: 2-3

2. Strain gauge measurement:

   - JP401: close
   - JP402: 1-2
   - JP403: 1-2

By changing these connections, the wiring diagram is modified. An essential part of this connection is the 2.5 V reference voltage source. For this purpose, the circuit MCP1525 (IC401) is used, which achieves an accuracy of $\pm1\%$, has a low-temperature drift (±50 ppm/°C max), and an input voltage of 2.7–2.5 V. The MCP6271 circuit (U401) is an operational amplifier connected as a constant current source to supply the resistive bridge.

Ad 1) In the circuit for temperature measurement using the PT1000 sensor, the voltage from the output of the circuit U401 (switch JP403—position 2–3) is applied to the output terminal REF+. The REF- terminal is connected to zero potential through the precision resistor F404. In the circuit diagram, resistor F404 is connected to the BPDSW pin of the used ADC lead. However, when the converter is active, the BPDSW pin is programmatically connected to zero potential. The replacement wiring diagram is presented in Fig. 2.30. In a 4-wire RTD configuration, two wires link the sensing element to the monitoring device on both sides of the sensing element. One set of wires delivers the current used for measurement, while the other set measures the voltage drop across the resistor. With the 4-wire configuration, the instrument will pass a constant current (I) through the outer leads, REF+ and REF−.

The parameters of the PT100 thermometer are the following:

- 4-conductor design
- $R = 1000\,\Omega$, at $t = 0\,°C$
- Measuring range: -50 to $205\,°C$
- The time constant—the time duration for the sensor to reach the correct temperature when the temperature of the medium changes suddenly

  - air environment: $\tau = 17$ s
  - fluid environment: $\tau = 135$ s

The relation (Eqs. 2.13 and 2.14) was used to calculate the temperature value. There are also available tables with the calculated coefficients in the data-sheet of the thermometer itself.

Fig. 2.30 The replacement wiring diagram with PT1000 sensor

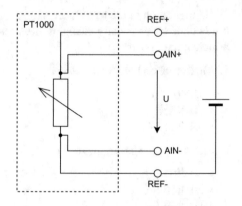

$$R = 1000(1 + At + Bt^2 + C(t - 100)t^3) \tag{2.13}$$

within the temperature range—50–0°C.

$$R = 1000(1 + At + Bt^2) \tag{2.14}$$

within the temperature range of 0 to 600°C,
where

$A = 3.9083 * 10^{-3} [°C^{-1}]$
$B = -5775. * 10^{-7} [°C^{-2}]$
$C = -4183. * 10^{-12} [°C^{-4}]$

Ad 2) When connecting for measuring on a strain gauge, the main difference is in the connection of the REF+ and REF− terminals. The constant current source U401 is not used in this connection. The REF+ terminal is connected to REFIN(+) and to the 3.3 V supply voltage (switch JP402, JP403: 1-2). The REF- terminal is connected to the REFIN(−) pin (switch JP401—connected) and via the BPDSW pin to zero potential when the transducer is activated. The terminals AIN+ and AIN− are the measuring terminals.

Software Control of the ADC Converter Parameters The parameters of the ADC converter can be changed using its three pins: GAIN, FILTER, and PDRST. These parameters are set in the firmware for a specific solution. The connection diagram of the MCU itself and the peripherals is shown in Fig. 2.31.

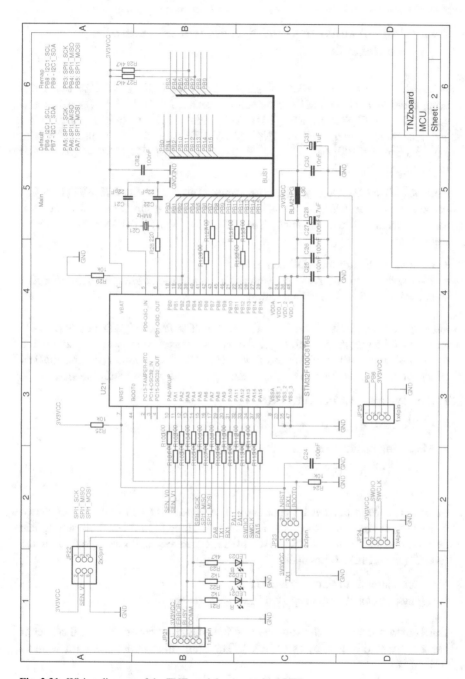

Fig. 2.31 Wiring diagram of the TNZ module—part with MCU

2.4.1 Implementation of the Specific uBUS Protocol Commands

A specific feature of the TNZ measurement module is the possibility of configuring each channel. The MCU pins control GAIN, FILTER, and PDRST parameters. The CMD_SLAVE_SET_PARAM (Table 1.29) function is used to set and load these parameters, or CMD_SLAVE_GET_PARAM (Table 1.30) with parameter names PARAM_AD_GAIN (0x10), PARAM_AD_FILTER (0x11) and PARAM_AD_POWER (0x12).

Activation/Deactivation of Measurements The CMD_SLAVE_SET_PARAM function, with the PARAM_AD_POWER parameter, is implemented to activate measurements on the channel. The CMD_SLAVE_SET_PARAMtaparameter values are:

- 0—deactivation of the ADC converter.
- 1—activation of the ADC converter. The automatic reading of data will start according to the currently set parameters.

Setting the Sensitivity of the ADC Converter The PARAM_AD_GAIN parameter needs to be modified according to the type of measurement, which is configured using HW links JP401m JP402 and JP403 (Fig. 2.29). After modifying the physical connection, it is necessary to set the correct gain value of the ADC converter:

- Strain gauge measurement: GAIN = 1

  – PARAM_AD_GAIN = 1

- Measuring on the sensor PT100: GAIN = 128

  – PARAM_AD_GAIN = 0

Setting the Measurement Frequency The used A/D converter supports two measurement frequencies, or data readings: 10 Hz and 16.6 Hz. This setting can be modified using the PARAM_AD_FILTER parameter. Possible values of the PARAM_AD_FILTER parameter:

- 0—measurement frequency is 10 Hz.
- 1—measurement frequency 16.6 Hz.

Data Format from the Measurement on the ADC Converter The used converter has a word width of 3 bytes (24 bits). The read value is defined by a formula (Eq. 2.15)

$$Code = 2^{n-1} * \left[AIN \frac{GAIN}{V_{ref}} + 1 \right] \qquad (2.15)$$

where $n = 24$, GAIN is the gain set, and V_{ref} is the reference voltage value ($V_{ref} = 2.5$ V).

Function CMD_SLAVE_GET_VALUE returns 5 B, of which the first 4 B are the measured value (CODE value in 32-bit integer format), and the 5th byte is status information about the read value:

- Bit 7—ready bit. 0—conversion is available
- Bit 6—filter bit. 1–10 Hz filter is selected, 0–16.6 Hz filter is selected
- Bit 5—error bit. 1—An error occurs during the conversion
- Bit 4 to 3—ID bits. Indicates the ID number for the AD7780. Value is 01
- Bit 2—gain bit. 1—gain $= 1$, 0—gain $= 128$
- Bit 0 to 1—status pattern bits. 00—serial transfer from the ADC was performed correctly

2.4.2   Implementation of Sensors for the TNZ Measuring Module

Special sensors AGS-01 and AGS-02 were developed for the TNZ measuring module for monitoring deformations of solid substances. In this particular case, the subject is a deformation of the asphalt body of the road in the horizontal and vertical directions. During the development of the described sensors, great emphasis was placed on the mechanical design, as the described sensors must withstand mechanical stress and temperature during their installation, which is carried out during the asphalt laying. Asphalt has a temperature of 150–$200°$ C during the laying. The proposed sensors are placed in the asphalt body of the road and must withstand extreme mechanical stress during the asphalt compacting and finishing. The designed structures can withstand these conditions for the time necessary for installation, which is about 30 min.

Sensor for Detection of the Longitudinal Deformation—AGS-01 Sensor AGS-01 (Fig. 2.32) consists of a longitudinal sensor body on which a strain gauge is installed and two arms that ensure mechanical deformation of the sensor body when force is applied to these arms. The body of the sensor itself can be subjected to tension or pressure. The distance between the arms is from 12 to 15 cm, and it can be changed by turning one of the arms located on the screw-plate. The internal design of the sensor body ensures safe storage of the cabling to the strain gauge.

The material ERTALON 66 SA[9] was used for manufacturing the sensor body. The properties of this material are higher mechanical strength and resistance to the influence of elevated temperatures. The material is suitable for machining on automatic lathes. The basic properties of this material are given in Table 2.4.

---

[9] https://www.eppplasty.cz/pdf/PA66.pdf.

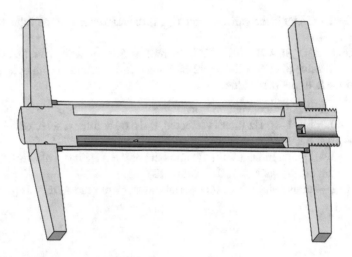

Fig. 2.32 Longitudinal deformation sensor model—AGS

Table 2.4 Basic properties of the ERTALON 66 SA material for manufacturing the AGS sensor

Property	Value	Note
Density	1.14 g/cm^3	
Water absorption	2.4%	In air at 23 °C and 50% relative humidity
Melting point	255 °C	
Thermal conductivity	0.28 W/(K.m)	at 23 °C
Operating temperature	−30°C to 180 °C	The maximum temperature is defined as short-term—in the order of hours.
Elasticity modulus	1650 MPa	

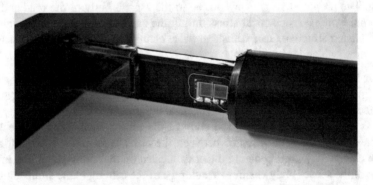

Fig. 2.33 Detail of strain gauge in the AGS sensor

Figure 2.33 shows a detail of the sensor AGS-01 body with strain gauge installed.

Sensor for Detection of the Transverse Deformation—AGS-02 The AGS-02 sensor (Fig. 2.34) consists of a vertical sensor body on which the strain gauge is installed, as well as the two load distribution plates in the shape of a circle, which

Fig. 2.34 The transverse deformation sensor

ensure mechanical deformation of the sensor body when the force is acting. The placement of the strain gauges is similar to that of the AGS-01 sensor. The sensor material is ERTALON 66 SA, as well.

2.5 Sensory Modules

Sensory modules represent specially designed solutions for use in specific conditions. Those are the different modular solutions that can cooperate with other modules than it was primarily intended for. The two solutions are presented in the following part:

- The OWS module (One-Wire Slave) solution for using any sensor on an 1-wire bus. It primarily cooperates with the OWB measuring module.
- The RTM module (Radio Transmitter Module). Sensory module with support for sending the measured data via the radio interface. It primarily cooperates with the RRM measurement module.

2.5.1 The OWS Sensory Module

For the 1-wire bus, there are slave modules that have the function of the ROM memory (e.g., DS2401), PROM memory (e.g., DS28EC20), RAM (e.g., DS2423), and digital thermometer (e.g., DS18B20). There is no sensor other than the temperature sensor for the 1-wire bus. Due to the narrowly limited availability of devices for

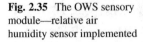

Fig. 2.35 The OWS sensory
module—relative air
humidity sensor implemented

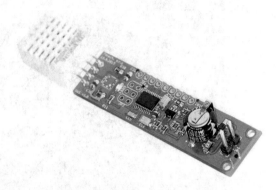

the 1-wire bus, a universal device intended for the 1-wire bus was designed and
implemented in the form of a configurable and easily modifiable solution based on a
microcontroller from the STM32 family, specifically STM32F030 and STM32L031.
The OWS (One-Wire Slave) module—Fig. 2.35 is a universal measuring module
communicating via a 1-wire bus.

The OWS module is a 1-wire bus slave device that can contain any sensor. The
main features of the OWS module are:

- It does not require an external power supply or a battery; power is provided from
  the 1-wire bus when there is no communication on the bus.
- It allows the connection of any sensor that can be connected to the peripherals of
  the used microcontroller: SPI, I2C, UART, GPIO.
- It supports the standard transmission rate of 16.3 kbps.
- In a completely discharged state, it takes approximately 30 s to bring it to full
  functionality.
- When discharged to a value close to 0 V, the module is ready in seconds.
- Unlike the conventional 1-wire bus controllers, the OWS device supports the
  supercapacitor fast charging function in cooperation with the OWB measuring
  board.

Since the OWS module represents a universal 1-wire bus slave module, each
module has information about which sensor it contains and other additional values,
which are the name of the sensor and its location. This additional data was added
for easier sensor identification in the final installation. The functionality of the OWS
sensor module was tested using the OWB measurement module. A 200 m long bus,
10 OWS sensor modules, and 20 DS18B0 sensors were used in the test connection.
The OWS modules worked as expected, and no communication errors were noted.

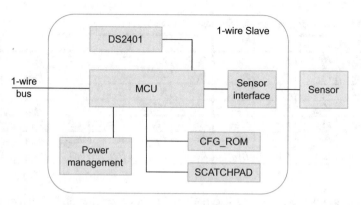

Fig. 2.36 OWS internal structure [11]

2.5.1.1 Application Program Interface of the OWS Module[10]

Each 1-wire bus device must implement basic commands to address and read/write data. Those commands have a 1-byte value and are denoted by a symbolic entry. For addressing purposes, these are READ_ROM (0x33), SEARCH_ROM (0x55), MATCH_ROM (0xF0), and SKIP_ROM (0xCC) commands. For the data handling, these are COPY_SCRATCHPAD (0x48), READ_SCRATCHPAD (0xBE), READ_MEMORY (0xF0), and CONVERT (0x44) commands. The functionality of these commands is explained in the further section.

The principal design of the OWS functionality is shown in Fig. 2.36. The OWS contains the DS2401—a Maxim Integrated 64-bit identifier. The OWS acts as a standard DS2401 identifier with the base set of commands. Advanced OWS properties for communicating with the connected sensor and reading and transferring the measured value are implemented by extending the set of supported functions to standard features such as READ_SCRATCHPAD or CONVERT [6] that the original DS2401 identifier does not recognize.

The OWS operates in the parasitic power mode. That is, it is powered directly from the 1-wire bus. It is powered using power management circuitry, including a supercapacitor, during communication and measurement. A 700 ms period is used to recharge the supercapacitor serving as a temporary power source to secure a large enough power reserve. It is impossible to communicate over the 1-wire bus during the supercapacitor recharging period, as the bus is used to power up the supercapacitor. The DS18B20 module defines the time required for the temperature conversion depending on the resolution. From 94 ms at a 9-bit resolution to 750 ms at a 12-bit resolution. This time is, therefore, variable. For the OWS module, the 700 ms time includes the supercapacitor charging time and the conversion of

---

[10] This section describes the authors' solution, which was in the past partially published in Journal of Sensors—[11].

Table 2.5 Format of a data frame with measured values [11]

Type-byte	0	1	2	3	4	5	6	7	8
Humidity	H_1	H_0	T_1	T_0	X	X	0x1	V_{bus}	CRC8
Pressure	P_3	P_2	P_1	P_0	X	X	0x2	V_{bus}	CRC8
Sunlight	L_3	L_2	L_1	L_0	X	X	0x3	V_{bus}	CRC8
Voltage	$V1_2$	$V1_1$	$V1_0$	$V2_2$	$V2_1$	$V2_0$	0x4	V_{bus}	CRC8
Current	$I1_2$	$I1_1$	$I1_0$	$I2_2$	$I2_1$	$I2_0$	0x5	V_{bus}	CRC8
Binary	B_1	B_2	B_3	B_4	B_5	B_4	0x6	V_{bus}	CRC8
Resistance	$R1_2$	$R1_1$	$R1_0$	$R2_2$	$R2_1$	$R2_0$	0x7	V_{bus}	CRC8

the required physical quantity itself. Charging is completed after 700 ms, and the conversion is started. The result of the conversion is stored in the SCRATCHPAD memory. The conversion length varies and depends on the type of connected sensor and its specifications. Typically, it is in the order of tens of milliseconds. The READ_SCRATCHPAD command reads 9 bytes of the measured value from the SCRATCHPAD area and sends it to the 1-wire bus to retrieve the measured value. The frame format for the OWS solution is described in Table 2.5. The selected 9-byte data frame length was based on the DS18B20 standard temperature sensor data frame.

The data frame formats for various measured physical quantities are displayed in Table 2.5: relative air humidity, atmospheric pressure, solar radiation intensity, electrical voltage, electrical current, electrical resistance, and binary inputs. The data frame for the relative humidity measurement contains the measured humidity value in the first 2 bytes, and the other 2 bytes represent the temperature. In the data frame for measuring the atmospheric pressure and solar radiation intensity, the first 4 bytes represent the measured value. When using a sensor that measures electrical voltage, electrical current, and electrical resistance, it is possible to implement two 24-bit converters—the frame contains up to 2 values, each of 3 B. The data frame for binary values allows reading 6 bytes and 48-bit information, respectively. The 6th byte in each frame sequence determines the type of data frame. The V_{bus} item represents the value of the OWS power supply voltage. In specific cases, triggering the measurement may be energy intensive, so it is necessary to check the power voltage level of the entire OWS. The last item is the CRC8 checksum.

Universal 1-Wire Module Design All existing 1-wire slaves can operate in parasitic power mode, in which they, in the IDLE state, use the bus potential to charge the supercapacitors, which then serve as a power supply during communication. Supercapacitor refers to a capacitor with sufficient capacity, which can supply power to the slave for a sufficiently long time when the slave proceeds with communication or measurement. In Fig. 2.37, it is labeled as SuperCap. The same principle is used in the design of the OWS. From a power point of view, the OWS has to meet the following prerequisites:

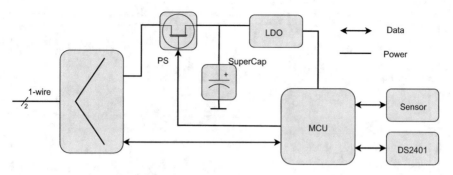

Fig. 2.37 Principal diagram of the hardware design [11]

1. If the capacitor was discharged and the OWS was connected to an 1-wire bus, it has to be charged without the intervention of the control logic and the control microcontroller, respectively.
2. Once the minimum supply voltage necessary to run the control microcontroller is reached, the process of charging is taken over by the microcontroller logic.
3. If an active level (logical 0) is detected on the bus, the charging of the capacitor is immediately terminated, and the OWS switches to the active mode in which it communicates with the 1-wire bus master.
4. When the communication is over, the OWS switches to the reduced power mode and activates the charging of the capacitor.

The proposed OWS consists of a sensor that measures the desired quantity (humidity, pressure, light intensity, and others), a control microcontroller, a 1-wire bus controller, power supply management of the entire OWS, and a capacitor in the role of a power source. It is necessary to minimize the power consumption of the entire equipment to ensure the greatest possible reliability. This is achieved by using the power-saving modes of the control microcontroller. The OWS may be found in active mode and standby mode. In active mode, the current consumption can be reduced by decreasing the clock frequency of the control microcontroller. In the passive mode, the consumption is minimized by turning off the unnecessary peripherals and switching the microcontroller into standby mode.

Hardware Design In Fig. 2.38 is shown the wiring diagram of the part that ensures the power supply of the OWS module. The 1-wire bus is represented by the X1 connector (Fig. 2.38, part D2). For writing to the bus, a signal marked 1WW is used. It can change the level of the 1-wire bus with the help of transistor Q1. The 1WW signal is connected to a pin PA1 of the used microcontroller. The supercapacitor C1 (Fig. 2.38, part A2) is used as a voltage source, which is connected to the voltage regulator (Low Quiescent Current LDO Regulator) with the low current consumption MCP1703 (IC3, part A3).

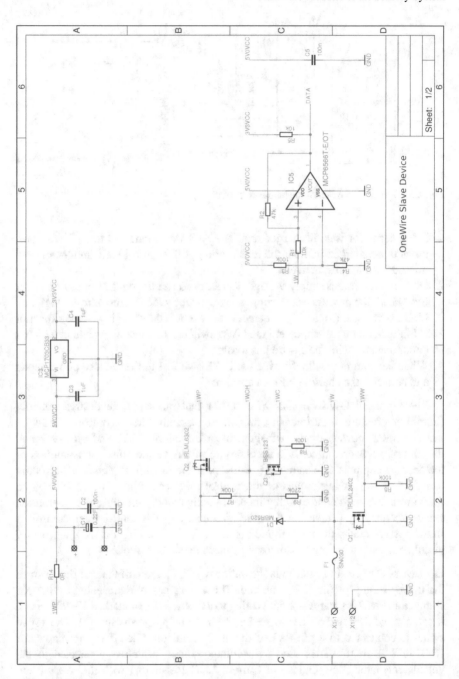

Fig. 2.38 OWS, 1-wire part

The microcontroller controls the supercapacitor charging when no communication is in progress. According to the specification, the OWS has to begin charging the capacitor after being connected to the 1-wire bus without any intervention of the control electronics. This condition occurs when the capacitor is completely discharged. The microcontroller is inactive after connecting the OWS module with discharged supercapacitor to the bus. After reaching the minimum voltage on the supercapacitor required for the control microcontroller, the controlled charging is activated. The switch (Fig. 2.38, transistor Q2, and Q3) is switched on, allowing charging at the maximum speed. The microcontroller's internal A/D converter is used to monitor supercapacitor voltage levels to reduce the total power consumption of the entire device. When the voltage on the supercapacitor is not high enough to start the microcontroller, the supercapacitor charging is controlled by an analog circuit based on the voltage divider principle.

The input signal of the 1-wire bus, marked as 1W, is fed to the input of the low-Power Open-Drain Output Comparator MCP6566 (circuit IC5). The output from the comparator (DATA signal) is fed to the microcontroller pin PA5, where it is subsequently processed.

Power and OWS Start-Up   The OWS does not need any external power source for its operation. The communication bus voltage in the parasitic power mode (parasite power) is used as a power supply in the IDLE mode. An essential part of the OWS hardware solution was to design a way to control the supercapacitor charging. With a fully discharged supercapacitor, the OWS has to start charging automatically until the voltage reaches the level at which the microcontroller will start. After reaching this voltage level, the microcontroller takes over the charging process. In Fig. 2.39, the charging current is 12 mA in the first approximately 17 s. If the supercapacitor voltage exceeds the threshold level at which the microcontroller starts operating, the controlled supercapacitor charging is switched on. At that point, the charging current reaches 80 mA, caused by switching on the PS switch itself. The current drops with exponential dependence and stays at 330 mA, which is the consumption of the microcontroller and additional circuitry in the STOP mode. The value of the supercapacitor capacity was 330 mF. The graphs (Fig. 2.39) are for the OWS module with the STM32F030 microcontroller. When using a low-power microcontroller, the graph differs by significant shortening of the first phase of the supercapacitor charging (part with the constant current consumption) to approximately 1 s.

The second part of the wiring diagram, shown in Fig. 2.40, represents the connection of STM32F0 microcontroller and peripherals. The OWS module has an implemented monitoring of the supply voltage, or voltage on the supercapacitor (Fig. 2.40, part A2). In the case of the low supply voltage, communication with the OWS module is not allowed, and at the same time, the 1WC signal is activated, which achieves fast charging of the supercapacitor.

One of the requirements was to design the OWS to allow the connection of standardized sensors. Under the "standardized," the sensor communication interface is understood. It is possible to connect the following to the OWS:

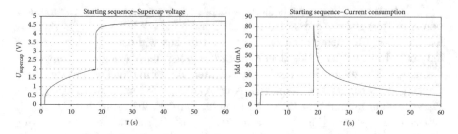

Fig. 2.39 Current consumption and voltage on supercapacitor during the starting sequence. STM32F030 microcontroller was used [11]

- Sensors communicating using the I2C interface, such as a sensor of visible and UV light Si1145 and atmospheric pressure sensor BMP180
- Sensors communicating using the SPI interface, such as a 24-bit A/D converter AD7195
- Sensors communicate using the I/O interface of the microcontroller ports, such as AM2302 humidity sensor, which uses its own serial communication protocol

Design versatility is ensured already in the firmware design. When adding a new sensor, it is necessary to implement SensorInterface_t interface (Listing 2.2).

```
typedef struct{
    void (*Init)(void* interface, void* settings);
    void (*Measure)(void);
    void (*Read)(void);
    int  (*IsReady)(void);
    void (*PowerDown)(void);
    void (*PowerUp)(void);
}SensorInterface_t;
```

Listing 2.2 Interface to implement new sensor in OWS module

Several functions ensure the proper functionality.

- Init()—initialization procedure for a particular sensor. Parameters to this function are specific hardware peripherals (I2C, UART, Digital Pins) and additional settings for selected peripherals.
- Measure()—starts the measuring procedure. The measurement procedure may be more complicated since, in several cases, it included a few readings of sensor values, which must be a certain delay. The LP_Delay() (low-power delay) function, used at these waiting delays, was created to ensure a minimum power consumption. The Measure() function must be implemented at the level of the OWS application itself. This function stops the supercapacitor charging and measures the sensor at the correct time points.
- Read()—this function ensures the correct formatting of the response data following Table 2.5.
- IsReady()—helper function that informs about valid measured data.

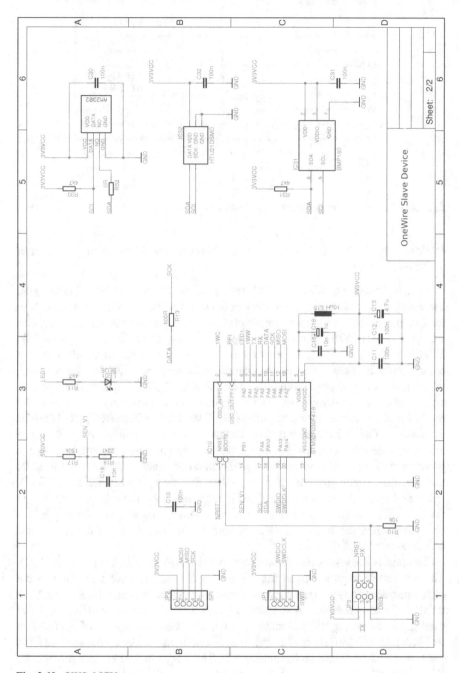

Fig. 2.40 OWS, MCU part

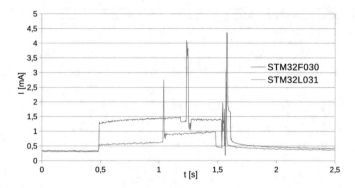

Fig. 2.41 Current consumption comparison in the RUN mode [11]

- `PowerUp()` / `PowerDown()`—these functions ensure power management of the used sensor.

In the implementation, two microcontrollers were selected: STM32F030 belonging to general-purpose microcontrollers, and STM32L031, which belongs to the low-power category. The charging process is similar for both microcontrollers, except that the STM32L031 can boot at a lower supply voltage than the STM32F030. In Fig. 2.41, a comparison of the selected microcontrollers' power consumption is shown. As test commands were used the CONVERT command, which starts the sensor measurement, and the READ_SCRATCHPAD command, which reads the measured value. In the beginning, the microcontroller is in STOP mode, and the communication begins at 0.5 s. The OWS switches to the RUN mode. Processing CONVERT command is as follows: since this command may also come at a time when the supercapacitor does not have enough energy to provide a reliable power supply for a period of 600 ms, the microcontroller switches to LP_RUN and activates supercapacitor charging. After this time, it switches to the RUN mode and reads the sensor value. This is represented by the first peak in Figure OWSCURRENT. The second peak is the processing of the READ_SCRATCHPAD command when the measured value is sent to the bus. This is followed by the transition to the STOP mode.

The power consumption of the OWS in the IDLE state is 330 μA (Fig. 2.41). This current is supplied from the 1-wire bus in regular operation. During the communication, this current rises to 1 mA with STM32L0 and 1.5 mA with the STM32F0 microcontroller. A supercapacitor of 330 mF capacity is used as a power source. The following test was aimed at measuring the consumption of the OWS in its IDLE and ACTIVE states. The maximum current that can be used is derived from the 1-wire bus controller on the master side. The 1-wire master bus can provide the short-term 200 mA current. The master module cyclically measures the 1-wire bus current; no communication is allowed when the 25 mA is exceeded. It is supposed that one of the connected supercapacitors is recharging.

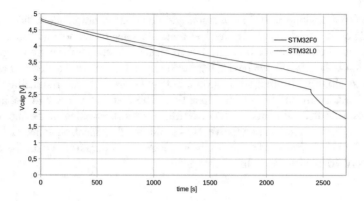

Fig. 2.42 Supercap voltage dropout in the STOP mode [11]

Discharging Test The tests were always based on the assumption that the supercapacitor is fully charged. The test began after disconnecting the OWS of the 1-wire bus. Figure 2.42 shows the voltage state of the OWS supercapacitor (Vcap) in the STOP mode, that is, at the time when there is no communication on the bus. The initial voltage value was $Vcap = 4.8$ V. The test was performed for 45 min.

V_{cap} voltage drops rapidly after 40 min when using an STM32F0 microcontroller. According to [12], the minimal supply voltage of this microcontroller is $Vdd = 2.4$ V; below this level, the microcontroller goes into a permanent reset state. The STM32L0 microcontroller belongs to the "ultra-low power" category. The minimal supply voltage of this microcontroller is $Vdd = 1.8$ V [13], which allows it to operate reliably even at lower voltages. This test focused on maintaining the required voltage on the supercapacitor, even when disconnected from the 1-wire bus. After disconnecting the OWS from the bus, the module is in the STOP state. If the OWS reconnects to the 1-wire bus with a nonzero voltage level, its charging will be much faster.

The proposed solution consists of the control microcontroller STM32F030F4 or STM32L031F6, a supercapacitor that serves as a power supply, a 1-wire bus controller, and a sensor that measures the required physical quantity. During the design and implementation phase, a great emphasis was placed on the low power consumption of the entire module. Low-power consumption is needed due to the modules' use: dozens of sensors can be connected to the 1-wire bus. The more energy-efficient solution uses the STM32L031 microcontroller since, in the RUN mode, it has 33.62% lower power consumption than the one with the STM32F030 microcontroller. The uniqueness of this solution lies in the exact implementation of the universal 1-wire module of the 1-wire bus. For different sensor types, other than a temperature sensor, there is no solution like a 1-wire slave. There was no 1-wire protocol modification in implementation, nor any new family code or new 1-wire function codes were added. The solution is fully compatible with the 1-wire standard.

The OWS contains a ROM that stores additional information about the measurement type, the measurement's physical unit, the sensor's name, and the limits of the measured physical quantity. The 1-wire master uses this information to provide a detailed overview of the 1-wire bus sensors. The OWS was tested in real-time operation along with other sensors on a 500 m 1-wire bus. After successful tests, OWSs were permanently put into operation. There are 15 OWS with relative air humidity sensors, with another 100 DS18B20 sensors installed in the monitoring system of temperature and relative air humidity of production halls and office spaces. This solution works flawlessly in continuous operation, where the data is read every 10 min for more than one year.

2.5.2   The RTM Sensory Module

The motivation behind the design of the RTM (Radio Transmitter Module) module was to create a sensor module that can be used in situations where it is not possible to use a bus solution. A typical example of use is temperature monitoring in stored bulk materials. The RTM sensor module—Fig. 2.43 is an autonomous module with one-way RF communication, containing up to 5 different sensors. The RTM module cooperates with the RRM measurement module, which receives and processes the data sent by the RTM module. The main features of the RTM module are:

- It allows the connection of any sensor that can be connected to the peripherals of the used microcontroller: SPI, I2C, UART, GPIO.
- The baud rate is 1000 baud.
- One AA battery is used as a power source.
- The data-sending interval is 5 min at a temperature of 20 °C. This interval is parameterized by the value of the measured temperature and varies at a rate of 8 s/°C.
- The simple design of the device cover enables service and battery replacement.

In the long-term tests of the RTM sensor module, the battery life, as an energy source, was 9 months to 1 year. This was achieved by using the microcontroller's

Fig. 2.43 The RTM sensory module

power-saving mode, which disconnects all the peripherals, including the microcontroller's core, when it is not measuring. The real-time circuit, which is a part of the used microcontroller, secures the transition to the operating mode again.

The basis of the hardware design of the RTM module was the STM32F030 microcontroller in the TSOP20 package. The mechanical design of the RTM module consists of three parts:

- Battery—the case for holding the battery is located on the back of the printed circuit board.
- Control part—it consists of a printed circuit board on which there is an MCU and auxiliary circuits.
- The RF signal transmitter module is connected to the control part using a connector. It contains a transmitter and an antenna.

The hardware design of the RTM module is shown in Fig. 2.44. The power source is one AA battery. With the help of the used step-up converter NCP1402SN50T1; the output voltage VOUT=5V is created from this voltage. The basic properties of this converter are:

- Extremely Low Start-up Voltage of 0.8 V
- Operation Down to Less than 0.3 V
- Output Voltage Accuracy ±2.5%
- Low Converter Ripple with Typical 30 mV
- Output current = 130 mA
- Output voltage = 5 V

The 5 V voltage is used to power the transmitter module. The other components are powered by a voltage of 3.3 V, which is provided by the LDO regulator MCP1703 with an ultra-low self-consumption (2 mA) and an output current of up to 250 mA. The printed circuit board's physical design was implemented so that various sensors could be used. The SPI bus terminal (Fig. 2.44 connector JP2), serial port—connector JP3, and programming interface—connector JP1 are prepared on the board. The I2C interface connects the LM75AD temperature sensor, whose measured temperature range is -55 to $125\,°C$ and the resolution is $0.125\,°C$. This sensor is in the SO8 housing. The module contains two signaling LEDs that indicate the current status (active mode, data-sending signal).

2.5.2.1 Application Communication Protocol for the RTM and RRM Modules

The proposed monitoring system will work with one-way communication. For the data packet, the following requirements were defined:

- Sufficiently large address space for sensors
- Variable packet items
- Content of the data part: telemetry data and data from connected sensors

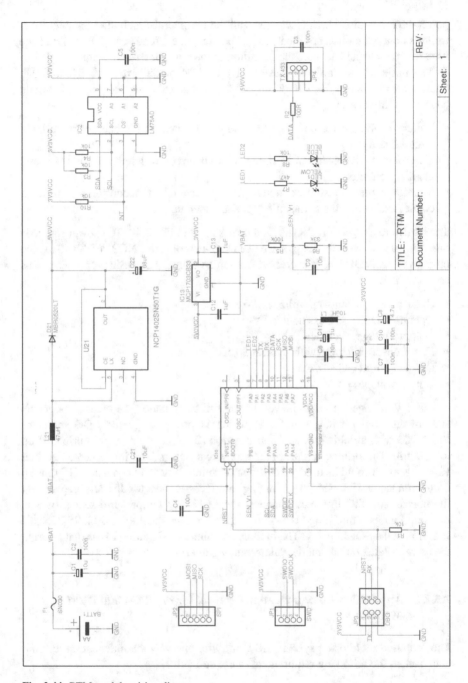

Fig. 2.44 RTM module wiring diagram

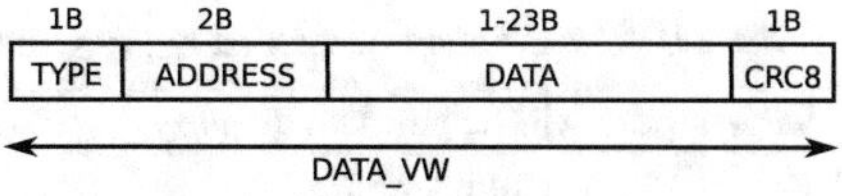

Fig. 2.45 Application packet format [9]

1B	2B	1-23B	1B
TYPE	ADDRESS	DATA	CRC8

DATA_VW

Fig. 2.46 The meaning of the bits in the TYPE byte [9]

Byte	0							
Group	Content					Version		
Bit	7	6	5	4	3	2	1	0
Info	V_1	V_2	V_3	V_4	V_5	R_1	R_2	R_3

- Packet check
- Minimalistic demands for overhead communication

The RTM broadcast sensor module uses ASK modulation. Since this modulation method is sensitive to the repeating bit values, coding is used before the modulation itself, which creates two hexadecimal bits from one byte (8 bits), ensuring that ones and zeros will alternate. The coding principle can be expressed formally using the relation in Eq. 2.16:

$$d_i^0 d_i^1 = encode(data_i) \tag{2.16}$$

where $data_i$ is the i-th coded byte and d^0 and d^1 are the 6-bit bit groups. After encoding the content, a synchronization part, start symbol, and frame length information are added to the packet, then the encoded data part is inserted, and a checksum is added. For a message that is 64 bits (8 bytes), the resulting communication frame will have a data portion of 96 bits (8 * 12 bits) and an overhead of 84 bits; thus, a total of 180 bits. The time required to send this packet at a transmission speed of 1000 baud will be 0.18 s.

The Communication Packet Format[11] Parts of the resulting communication packet are shown in Table 2.3, which is sent directly to the transmitter circuits. The payload portion of this packet, labeled DATA_VW, forms the content that the RTM module sends. The structure of this part is shown in Fig. 2.45. It contains parts TYPE (defines the format of the data part of the packet), ADDRESS (address of the RTM module), DATA (data from sensors), and CRC8 (checksum).

The address space for the sensor probes was defined in the width of 2^{16} of unique addresses. The packet data has a maximum length of 23 B. The last item is the CRC8 checksum using the $X^8 + X^5 + X^4 + 1$ polynomial, which is used in the 1-wire standard [14].

TYPE The first byte in the packet (Fig. 2.46) defines the entire packet's format and the list of quantities in the packet.

---

[11] This section describes the authors' solution, which was in the past partially published in Journal of Sensors—[9].

Table 2.6 The meaning of the first byte—content part, version 000 [9]

Value type	The physical quantity	Size	Range	Resolution	Unit
V_1	Temperature	2 B	-127–127	0.0125	°C
V_2	Humidity	1 B	0–100	0.5	N/A
V_3	Atmospheric pressure	2 B	0–8192	0.25	hPa
V_4	Light intensity	2 B	0–32,768	0.5	lx
V_5	UV index	1 B	0–32	0.25	N/A

TYPE byte is divided into two parts: Content and Version. The three bits $(R_1 R_2 R_3)$ define a version, respectively, meaning of Content. The individual bits V_i indicate whether there is a part in the packet that contains the measured value of a defined quantity. The data frame format breakdown by revision $(R_1 R_2 R_3)$ is:

- 000 environmental quantities
- 001 electrical quantities
- 010 universal A/D converter

Version 000—Version value (0,0,0) was defined for the probe measuring environmental variables. For this version, the meaning of the Content part is in Table 2.6:
Version 001—The data frame format for measuring electrical quantities.

- $V_1 V_2$—number of sets of measurements
- V_3—electric resistance, 2 B
- V_4—electric voltage, 2 B
- V_5—electric current, 2 B

Version 010—The data frame format for values from the n-channel analog-to-digital converter.

- $V_1 V_2$—number of converter bytes
- $V_3 V_4 V_5$—number of converter channels

ADDRESS Two bytes are reserved for addressing. The address space is from address 1 to 65535. There are no additional restrictions for the addresses defined.

DATA The length of the DATA part is given by the byte TYPE, in which is the DATA content defined. The first byte in the DATA section is always the telemetry data, respectively, battery level. All the other values are optional. In Fig. 2.47 are presented examples of the application frame for different Revisions.

Figure 2.47a example 1: Revision 000; data frame contains all the values TELE—telemetry, TEMP—temperature, HUM—relative air humidity, PRES—atmospheric pressure, LIGHT—light intensity, UV—UV index.

Figure 2.47b example 2: Revision 000; Content value in the TYPE configuration byte is 101000. DATA will only contain temperature data (TEMP) and atmospheric pressure (PRES).

a) Version = 000, example 1

Byte	0	1..2	3	4..5	6	7..8	9..10	11	12
Value	11111000	ADRR	TELE	TEMP	HUM	PRES	LIGHT	UV	CRC8

b) Version = 000, example 2

Byte	0	1..2	3	4..5	6..7	8
Value	10100000	ADRR	TELE	TEMP	PRES	CRC8

c) Version = 001

Byte	0	1..2	3	4..5	6..7	8..9	10..11	12
Value	01110001	ADRR	TELE	R1	U1	R2	U2	CRC8

d) Version = 010

Byte	0	1..2	3	4..6	7..9	10
Value	10001010	ADRR	TELE	ADC1	ADC2	CRC8

Fig. 2.47 Application frame examples [9]

Figure 2.47c: Revision 001; data frame for electrical quantities values. The number of sets of measurements is given by $V_1V_2 = 01$—in the data part, there will be two series of measurements. The triplet $V_3V_4V_5$ determines the measured values. Value 110 means that each set will contain the value of electrical resistance and electric voltage.

Figure 2.47d: revision 010; data frame for AD converter values. The word width of the AD converter is defined by $V_1V_2 = 10$. The width of the word will then be 3 B=24 bit. The triplet $V_3V_4V_5 = 001$ determines the channel number of the converter. Thus, there will be two values in the data section, each with a width of 3 B.

2.5.2.2   The RF Communication Security Proposal

An RRM module receives measured values from the independent wireless measuring probes (RTM) in the sensor system application. Communication between the probes and the measuring module is exclusively one-way—after sending the measured values, the measuring probes have no information on whether the sent message was received. The task of the measuring module is only to temporarily save the measured values from the received messages as long as the superior measuring station requests them. A 16-bit CRC sum is used to ensure the consistency of the received message. This CRC sum carries information about whether the data in the data part of the communication frame has not been modified.

The issue of secure data transmission is even more urgent in wireless transmission, especially in the field of radio frequency transmissions in the free band. Unlike Bluetooth or Wi-Fi technology, where the secure layer is a part of the given standard, radio frequency transmission technology does not define any security. It is also due to the very nature of the transfer itself—the format, coding, and encryption of the sent data packets depend on the application in which this transfer is made. Corrupting the data sent by the measuring probe is simple: receive the original message, decode it, change the value, encode it, and send it. Since communication is one-way, the receiving party cannot verify the source of this data. By adding secure communication, this problem can be eliminated. Since it is a one-way transfer, it is impossible to exchange encryption keys before the actual communication. We have defined the following requirements for the implementation of the secure transmission:

1. Use of a symmetric code
2. Computing-friendly encryption algorithm
3. Encryption algorithm suitable for the small lengths of source messages
4. Use of multiple encryption keys

Requirement 1 is related to the fact that this is a unidirectional transmission. Requirement 2 is justified since the probe module is based on a solution with a microcontroller in which the clock frequency was reduced due to longer battery life. As the data part of the communication frame can have a maximum length of 30 bytes, we require that the encrypted message has a minimum length (requirement 3). An important parameter when choosing an encryption algorithm is the encrypted block's length and the key's size. Given the minimum and maximum lengths of the communication frame, the appropriate block length is 8 B. At this length, a data frame can be 8 B, 16 B, and 24 B in length. In the final version, the sensor module will contain a set of encryption keys, from which it will choose one key by pseudo-random selection. The code that meets these requirements is the code SPECK [15].

2.5.2.3   Block Cipher SPECK and SIMON[12]

With the upcoming era of the Internet of Things and Pervasive Computing, there is a need to develop block ciphers with tight constraints such as area, power, memory, performance, throughput, and others. These are the so-called lightweight block ciphers intended explicitly for resource-constrained platforms. Lined up in the line is SIMON, a lightweight block cipher proposed by the NSA after prompting from the U.S. Government in 2013, along with the SPECK lightweight block cipher. The SIMON implementation on hardware has excellent results in terms of efficiency and is a powerful alternative to the existing AES [16]. The SPECK supports a variety of block and key sizes. A block is always two words, but the words may be 16, 24, 32, 48, or 64 bits in size. The corresponding key is 2, 3, or 4 words. The round function consists of two rotations, adding the right word to the left word, XORing the key into the left word, and then XORing the left word to the right word. The number of rounds depends on the parameters selected.

---

[12] This section describes the authors' solution, which was in the past partially published in Journal of Sensors—[9].

Table 2.7 The SPECK algorithm variants of block size and key size [9]

Variant	Block size (bits)	Key size (bits)	Rounds	(α, β)
1	$2 \times 16 = 32$	$4 \times 16 = 64$	22	(2,7)
2	$2 \times 24 = 48$	$3 \times 24 = 72$	22	(3,8)
3		$4 \times 24 = 96$	23	(3,8)
4	$2 \times 32 = 64$	$3 \times 32 = 96$	26	(3,8)
5		$4 \times 32 = 128$	27	(3,8)
6	$2 \times 48 = 96$	$2 \times 48 = 96$	28	(3,8)
7		$3 \times 48 = 144$	29	(3,8)
8	$2 \times 64 = 128$	$2 \times 64 = 128$	32	(3,8)
9		$3 \times 64 = 192$	33	(3,8)
10		$4 \times 64 = 256$	34	(3,8)

The SPECK has been optimized for performance in the software implementations, while its sister algorithm, Simon, has been optimized for hardware implementations. SPECK is an add-rotate-xor (ARX) cipher. The SPECK2n encryption maps a plain text of the two n-bit words (x0, y0) into a cipher text (xT, yT), using a sequence of T rounds. The key-dependent round function is defined as [15]:

$$R_k(x, y) = \left(\left(S^{-\alpha} x + y \right) \oplus k, \left(S^{\beta} x \oplus y \right) \oplus k \right) \tag{2.17}$$

where k is the round key, and rotation constants α and β are given in Table 2.7. Used operation are:

- $\oplus$ bitwise XOR
- Left circular shift, S^j, by j bits
- Right circular shift, S^{-j}, by j bits

The decryption rule is:

$$R_k^{-1}(x, y) = \left(\left(S^{\alpha} x \oplus k \right) - y, \left(S^{-\beta} x \oplus y \right) \right) \tag{2.18}$$

The SPECK key schedule reuses the round function to generate the round keys $k_0, \ldots, k_T$. The m-word master key $K = (l_{m-2}, \ldots, l_0, k_0)$ are used as follow [17]:

$$l_{i+m-1} = (k_i + S^{-\alpha} l_i) \oplus i \tag{2.19}$$

$$k_{i+1} = S^{\beta} k_i \oplus l_{i+m+1} \tag{2.20}$$

Figure 2.48 provides a schematic view on the round function and the key schedule of SPECK. R_i is the SPECK round function, with i acting as the round key. Differential cryptoanalysis can break 25 of 34 rounds of Speck128/256[13] with

---

[13] 128/256 denote block size of 128B and key size of 256B.

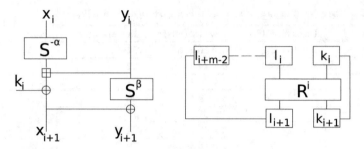

Fig. 2.48 The round function and the key schedule of SPECK [18]

$2^{253.35}$ time complexity using $2^{125.35}$ chosen plaintexts and 2^{22} bytes memory; or 23 of 32 rounds of Speck128/128 with $2^{125.35}$ time complexity and $2^{125.35}$ chosen plaintexts [17]. Distinguishers for reduced-round versions of Speck32/48/64 have been found by automated means, [18] and it is suspected that the same would happen to Speck128/256, given more computer power. According to European Network of Excellence in Cryptology stream cipher benchmarks (eBASC) [19], Speck is one of the fastest ciphers available, both for long as well as short messages and is comparable in speed to the stream cipher Salsa20. When implemented on 8-bit AVR microcontroller, Speck encryption with 64b blocks and 128b key consumes 192 bytes of Flash, temporary variables consume 112 bytes of RAM, and takes 164 cycles to encrypt each byte in the block [20].

From the available variants of the key length and the length of the encrypted block of the SPECK, variants were selected so that the means of the architecture of the used microcontroller were effectively used in their implementation. The STM32L0xx family of microcontrollers are 32-bit microcontrollers; the native word length is 32-bit. SPECK Variant 1 and Variant 5 were selected (Table 2.7). Variant 5 uses 128-bit (4 words) key length, and 64-bit (2 words) block size, therefore 8 bytes. The 8 bytes size is sufficient for a packet that contains measured temperature and humidity data. For packets longer than 8 Bytes, Variant 5 can be applied independently for each block of 8 bytes.

During the implementation of secure communication, requirements were defined for both the sending and receiving sides:

- Broadcasting side

  - It will contain a set of non-repeating encryption keys.
  - When sending data, a different encryption key is used than in the previous transfer.
  - The time for calculating the encrypted message must be minimal.

- Receiving side

  - The receiving party does not know the encryption key, but knows the set of keys from which the specific encryption key was selected.

- Based on the known format of the received decrypted message, it will be able to find the correct decryption key.

Transmitting Side The transmitting device will contain a set of keys, of which a quasi-random selection will select one key. In order to minimize the use of the limited memory capacity of the microcontroller, a key selection procedure has been proposed. This applies to Variant 5 (Table SPECK-T1). For variant 1, it will vary by word length: instead of 32-bit values, 16-bit values are used, and the key size will be a pair of sub-keys. Encryption key selection algorithm:

Algorithm SPECK

1. n random 32-bit numbers are generated and stored in the array of a size n. This is a set of partial keys: $K = k_1 k_2 k_3 \ldots k_n$
2. The random number j is selected; $0 \leq j \leq n - 1$
3. The key used for the encryption will be: $K_j = \kappa_j \kappa_{j+1} \kappa_{j+2} \kappa_{j+3}$, where $\kappa_{j+i} = k_{(j+i) \mod n}$

With the set size $n = 256$, we get 252 keys, each of a 128-bit size. The memory size used to store these keys is 256 words = 1024 Bytes. Another requirement for transmitting was changing the encryption key in subsequent communication.

The last criterion was minimizing the encrypted message computation time. The microcontroller is in the RUN mode, clocked at 8 MHz. At this frequency, the time required to apply the SPECK cipher on an 8-byte communication frame is 55 ms. For comparison, the calculation of the CRC (Fig. 2.45) for the same communication frame lasts 60 ms. Compared to the summary time, since the microcontroller switches to RUN state, over to the sensor measurement to RF transmission, the time needed to encrypt the message is negligible. When using the LM75 temperature sensor, the time slots are the following: time needed to initiate and get the sensor value $t_{init} = 200$ ms, the time to send the measured value over the RF interface at a baud rate of 1000 baud: $t_{transmit} = 30$ ms.

Receiving Side The receiving part uses the same microcontroller. Due to the nature of the application, only the RUN mode will be used, as there is no need to switch the receiving module into the low-power mode. In order to be able to decrypt the received message, a decryption key must be available. However, according to the transmitting module specification, one encryption key from set K (see Algorithm SPECK) is randomly selected for transmission. The receiving module must know this set of keys and find a key that decrypts the frame correctly. The algorithm was designed to determine the decryption key and decrypt the received packet (Fig. 2.49). Figure 2.49 uses the following labeling: P'—encrypted data packet, P— decrypted data packet, and K—set of available keys. The SPECK setup function prepares the sequence of the partial keys used for the decryption according to the

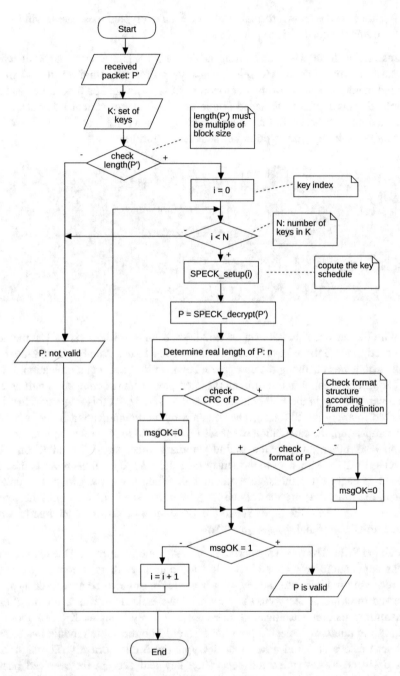

Fig. 2.49 Algorithm for determining the decryption key [9]

selected encryption key. When determining the correct decryption key, the following conditions must be met:

1. CRC8 of the received frame must be the same as the last byte in the received frame.
2. The length and format of the data part must correspond to the definition of the data part contained in the first frame of the communication frame.

For the N key set, it is necessary to run a decryption function maximum N-times and check the validity of the decrypted frame. When using the STM32L053 microcontroller with a maximum clock frequency of 32 MHz, the length of the encoded-word 8 Bytes, the calculation of key schedule value takes 131 ms, and decryption itself takes 19 ms. At the maximum number of 252 iterations, the decryption boundary time is 44.6 ms.

References

1. J. Ďuďák, M. Skovajsa, I. Sladek, Proposal of a communication protocol for smart sensory systems. In *Proceedings of the 16th International Conference on Mechatronics—Mechatronika 2014* (Brno, 2014), pp. 107–112
2. J. Ďuďák, G. Gašpar, S. Sedivy, P. Fabo, L. Pepucha, P. Tanuska, Serial communication protocol with enhanced properties–securing communication layer for smart sensors applications. IEEE Sensors J. **19**(1), 378–390 (2019). https://doi.org/10.1109/JSEN.2018.2874898
3. J. Ďuďák, The proposition of MODBUS protocol extensions, in *AIM 2010: Proceeding of 5th International Symposium. Advances in Mechatronics* (Trenčín, 2010), pp. 1–4. ISBN: 978-80-8075-469-3
4. J. Ďuďák, P. Cicak, CPN model of the MODBUS protocol, in *13th Mechatronika 2010* (Trencianske Teplice, Slovakia, 2010), pp. 43–45. ISBN: 978-1-4244-7962-7
5. Zetex Semiconductors, Low offset high-side current monitor. Datasheet ZXCT1022 (2008)
6. Maxim Integrated, Programmable resolution 1-wire digital thermometer. Datasheet DS18B20, Maxim Integrated Products, Inc. (2007)
7. J. Ďuďák, G. Gašpar, S. Sedivy, L. Pepucha, Z. Florkova, *Road Structural Elements Temperature Trends Diagnostics Using Sensory System of Own Design*, vol. 236 (2017). https://doi.org/10.1088/1757-899X/236/1/012036
8. G. Gašpar, J. Ďuďák, M. Behulova, M. Stremy, R. Budjac, S. Sedivy, B. Tomas, IoT-ready temperature probe for smart monitoring of forest roads. Appl. Sci. **12**(2) (2022). https://doi.org/10.3390/app12020743
9. J. Ďuďák, G. Gašpar, P. Tanuska, Implementation of secure communication via the rf module for data acquisition. J. Sensors **2019**(2) (2019). https://doi.org/10.1155/2019/7810709
10. M. Skovajsa, P. Fabo, L. Pepucha, I. Sladek, Proposal of a wireless measurement system for temperature monitoring of biological active materials, in *Advanced Mechatronics Solutions*, ed. by R. Jablonski, T. Brezina (Springer, Cham, 2016), pp. 367–372. ISBN: 978-3-319-23923-1
11. J. Ďuďák, P. Tanuska, G. Gašpar, P. Fabo, Arm-based universal 1-wire module solution. J. Sensors **5268247**(1), 16 (2018). https://doi.org/10.1155/2018/5268247
12. STMicroelectronics NV, Value-line Arm®-based 32-bit MCU with up to 256 KB Flash, timers, ADC, communication interfaces, 2.4-3.6 V operation. Technical report, STMicroelectronics (2021). http://www.st.com/resource/en/datasheet/stm32f030f4.pdf. Accessed 14 Nov 2022
13. STMicroelectronics NV, Access line ultra-low-power 32-bit MCU Arm®-based Cortex®-M0+, up to 32 KB Flash, 8 KB SRAM, 1 KB EEPROM, ADC. Technical report, STMicro-

electronics (2021). http://www.st.com/resource/en/datasheet/stm32l031f4.pdf. Accessed 1 Nov 2022
14. Analog Devices, Inc., Application note 27 - Understanding and Using Cyclic Redundancy Checks with Maxim 1-Wire and iButton Products. Technical report, Analog Devices, Inc. (2021). https://www.analog.com/en/technical-articles/understanding-and-using-cyclic-redundancy-checks-with-maxim-1wire-and-ibutton-products.html. Accessed 14 Nov 2022
15. G. Yang, B. Zhu, V. Suder, M.D. Aagaard, G. Gong, The simeck family of lightweight block ciphers, in *International Workshop on Cryptographic Hardware and Embedded Systems*, ed. by T. Güneysu and H. Handschuh (Springer, Berlin/Heidelberg, 2015)
16. R. Nithya, D. Kumar, Where AES is for internet, Simon could be for IoT. In *Procedia Technology* (2016), pp. 302–309. https://doi.org/10.1016/j.protcy.2016.08.111
17. L. Song, Z. Huang, Q. Yang, Automatic differential analysis of ARX block ciphers with application to speck and lea, in *Lecture Notes in Computer Science*, ed. by J. Liu, R. Steinfeld, vol. 9723. Information Security and Privacy (Springer, Cham, 2016), p. 50
18. Y. Liu, G. De Witte, A. Ranea, T. Ashur, Rotational-XOR cryptanalysis of reduced-round speck, in *IACR Cryptology ePrint Archive*, vol. 1036 (2017)
19. EBACS, *ECRYPT Benchmarking of Cryptographic Systems: ENCRYPT II. Measurements of Stream Ciphers, Indexed by Machine* (2017). https://bench.cr.yp.to/results-stream.html. Accessed 14 Nov 2022
20. R. Beaulieu, D. Shors, J. Smith, S. Treatman-Clark, B. Weeks, L. Wingers, The Simon and speck block ciphers on AVR 8-bit microcontrollers, in *Lightweight Cryptography for Security and Privacy*, ed. by T. Eisenbarth, E. Öztürk, vol. 8898. LightSec 2014 (Springer, Cham, 2014)

Chapter 3
Software Modules of the Sensory System

When software breathes life into hardware

An integral part of every measuring or sensory system is the software that ensures data collection, processing, and presentation. From the point of view of access to measured data, we can talk about a sensory system as an information system (IS). The described hardware modules form an interface to the IS through which the data is stored in the system. In the case of a sensory information system, we can think about parts of these systems in a broader context. In the presented solution, we called the software solutions nSoric suite. It is, therefore:

- Basic modules

  - **nSoric IS**—data model of the sensory system. The data model is built on a relational database. The MySQL database is used in the current implementation.
  - **nSoric senlib**—a library that provides basic functionality for all the other components. It contains functionality for communication with the nSoric API and locally connected measurement modules.

- Server parts

  - **nSoric API**—application programming interface for data access. This interface enables remote diagnostics and configuration of the measurement modules.
  - **nSoric Serve**—automated data collection. A system service ensures the reading of values at regular intervals and storing of the obtained data.

- Utility software for measurement modules

  - **nSoric Cofig**—configurator of the measurement modules. It allows the upload of information to the module, such as the module's name, date of installation, used version of firmware and hardware, list of sensors that the module contains, and parameters of these sensors, or the calibration coefficients.

© The Author(s), under exclusive license to Springer Nature Switzerland AG 2023
J. Ďuďák, G. Gašpar, *Design and Implementation of Sensory Solutions for Industrial Environment*, Signals and Communication Technology,
https://doi.org/10.1007/978-3-031-30152-0_3

- **nSoric Merula**—starting on-demand measurements on the measurement modules. The data is stored locally on the computer where the software is running.
- **nSoric Aurela**—visualization of sensory data by accessing the data stored on a remote server. It uses the nSoric API for communication.

During the development of the software part, emphasis was placed on multi-platform use and the robustness of the solution. Therefore, Java was chosen as the programming language. This language enables the creation of software in the role of a system service and an application with a modern graphical user interface. The resulting software works can be run on all the currently used operating systems (Linux, Mac, Windows).

3.1   Basic Software Modules

An information system is defined as a set of people, technical means (hardware, software), and data that ensure the required functionality and provide information for a defined and required purpose. Processes in the information system primarily include:

- Data collecting
- Data retention
- Data transfer
- Data processing
- Data and information provision

All these features of the IS are implemented in the basic software modules. Implementation of the database model ensures the preservation of data and its provision. The functionality of the nSoric senlib library enables data collection, transmission, and processing.

3.1.1   nSoric IS—Data Model of the Sensory System

When selecting a suitable platform for the IS implementation, we considered its extensibility, usability in similar solutions, and pricing policy. The primary criterion was a relational database. There are several big players on the market offering complex solutions. The portal db-engines.com[1] has published the "DB-Engines Ranking of Relational DBMS," which ranks database management systems according to their popularity. As of 10/2022, Oracle is the most used system,

---

[1] https://db-engines.com/en/ranking/relational+dbms.

followed by MySQL, Microsoft SQL Server, PostgreSQL, and IBM Db2. Due to its great extensibility, ease of availability, and license version, MySQL was selected. The IS model is divided into two parts: a model of user permissions and user membership to other entities and a model of the measurement application itself. When designing the system architecture, it was assumed that the system would have a centralized list of users and measurement installations (or applications)— Fig. 3.1. Part of the sensor solution is the storage of measured data in the data infrastructure provided by this IS. A separate database will be created for each sensor system installation (Fig. 3.2). These databases do not need to be on a single server physically but will be centrally managed using the first model that addresses user permissions. For unified access to the data from these systems, an application programming interface (API) has been created by which the end application can interact with the system.

The IS architecture is designed as a service to the end user—the measured data is stored in the cloud, which is a part of the sensory system. The end consumer (user) can access multiple installations, each with defined permissions. In the data model in Fig. 3.1, the purpose of the *Users, User Companies*, and *UserCompanies* tables is obvious. The *Apps* table represents information about a specific installation of the sensor system. Since the data from a specific installation may not be stored centrally on a single database server, the Servers' table contains information about the server on which the database for a specific application is created. In the *AppRestrictions* and *Restrictions* tables, restrictions are defined, such as restrictions

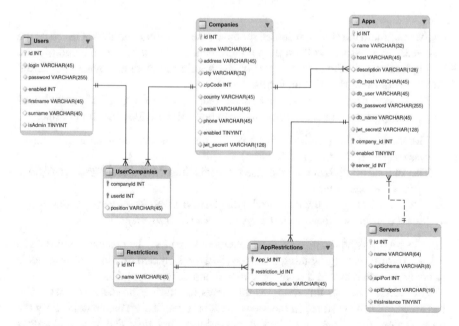

Fig. 3.1 Information system data model—part of application administration and authorization

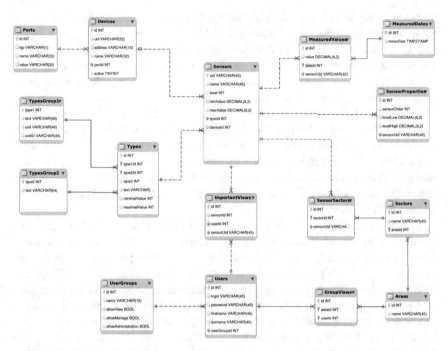

Fig. 3.2 Data model of the information system for environmental data collection—part of the measurement application

on the maximum number of installed modules, the maximum number of users, or the frequency of measurements on the measurement modules. Those constraints must be implemented in the software part of the nSoric server that provides the actual direct execution of the measurements.

The second part of the model is a model that represents a specific installation of the sensor system—Fig. 3.2. This model can be further divided into four parts:

1. The model of the physical wiring of the measurement modules in the sensor system (*Ports, Devices,* and *Sensors*)
2. The model of division of the used sensors into hierarchical groups (*Areas, Sectors, SensorSector*)
3. Model for storage of the measured data (*MeasuredValues, MeasuredDates*)
4. Model of user permissions in the system (*UserGroups, Users*)

The design followed the principles of data model creation; the presented model is in a 3-norm form. A programming interface (API) was created on top of these models to communicate between the database model and the end application.

An essential part of this model is the tables for storing the measured data. The measurement time is stored in the *MeasureDates* table. This table contains only the measurement time (*measDate*) and the record identifier (*id*). The measured values are stored in the *MeasuredValues* table. A record in this table consists of 3-part data (*value*, *dateId*, *sensorUid*), where value is the measured value itself, *dateId* is a reference to the measurement time in the *MeasureDates* table, and *sensorUid* is a reference to the sensor in the Sensors table.

3.1.2   nSoric senlib—Shared Sensor System Library

The nSoric senlib shared library is a core library providing the functionality to access measured data via the nSoric API application programming interface and direct access to measurement modules. The basic version of this library has been implemented in Java. There is an implementation in Python (nSoric pySenlib) and Go (nSoric goSenlib) for easy functionality implementation in other systems. The basic structure of the nSoric senlib library is the following:

- senlib.api—API client for nSoric API
- senlib.fw—ensures direct communication with measuring modules

In the standard case, the measured data is stored in a MySQL database. Data mirroring to the local database has been implemented in the library to eliminate repeated data downloading from the server. The SQLite was chosen as the local database due to its easy use—there is no need to install another DB server—the database uses the host computer's file system directly, as it is an entire relational database system—it is possible to make a local copy of a part of the remote database.

senlib.api The senlib.api package represents the API of the sensor system client. Figure 3.3 shows the principle of using the senlib.api library. An application accessing the measured data stored in the sensor system communicates with the senlib.api interface, which provides a choice of data source—locally mirrored data or data on a remote server. The application can also request access to measurement modules for diagnostics or measurement.

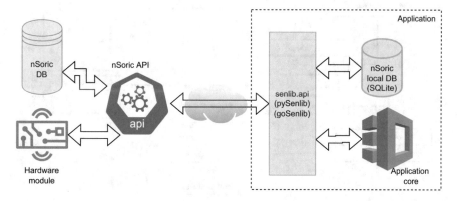

Fig. 3.3 Communication with a remote server

Content of the senlib.api package is:

- Communication interface using a secure SSL layer
- Object-relational mapping of all the tables
- Methods for local database synchronization. This is exclusively a one-way synchronization from a remote server
- Authentication process when communicating with the nSoric API

senlib.fw The senlib.fw package provides direct access to the measurement modules and their sensors. The classes' basic structure and dependencies are shown in Fig. 3.4. The measurement modules (e.g., OWB, TNZ, ...) are represented by the Device class. The Device class contains methods for essential communication with any measurement module implementing the uBUS protocol. The implemented methods can be divided into three groups:

- Connection methods:

    - *open()*—opening the connection to the measuring module
    - *close()*—closing the connection to the measuring module

- Information methods:

    - *getName()*—returns the name of the measurement module
    - *getFamily()*—returns the code of the measurement module: e.g., OWB, TNZ, RRM, ...

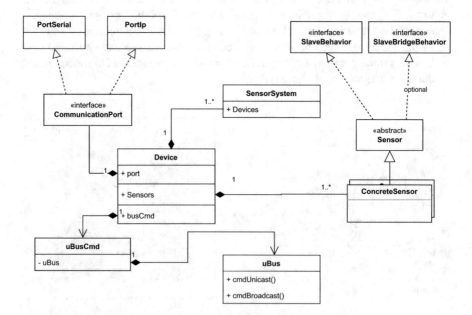

Fig. 3.4 Class diagram of s the senlib.fw package

- *getFW()*, *getHW()*—returns the firmware version and hardware solution of the measurement module
- *getSlaves()*—returns the list of active sensors
- *getSlave(offset)*—returns the instance of the requested sensor
- *getAddress()*, *getInfo()*, *getParam()*—other information methods

• Measuring methods

- *measureStart()* / *measureStop()*—starts/stops continuous measurement on sensors

Communication options are defined by the *CommunicationPort* interface. This interface defines the basic operations, such as *openPort()*, *closePort()*, and *send()* functions. The concrete implementation of these methods is in the *PortSerial* (serial port connection) classes and the *PortIp* (IP address connection) classes. In the case of connecting the measurement module via a physical RS-485 bus, the *PortSerial* class is used. The *PortIp* class is used if the measuring module is connected to the Internet, using the Serial/Ethernet or Serial/Wi-Fi converter. Those converters map the RS-485 serial interface to an IP address and a port. The uBusCmd class provides an implementation of the uBUS application protocol. The uBus class implements low-level communication—the actual formatting of the request, its sending and receiving.

The behavior of the sensor, connected to the measurement module, is defined by the SlaveBehavior interface. This interface defines the methods:

• *getValue()* / *setValue()*—reading/writing (for output submodules) of values
• *getParam(paramName)* / *setParam(paramName, paramValue)*—getting/writing the parameters of the given sensor
• *reset()*—software reset of the sensor
• *getState()*, *getType()*, *getName()*, *getLimits()*—other information methods

The Sensor abstract class implements the general methods of the *SlaveBehavior* interface. These are the ones whose implementation is the same for all the concrete sensors: *getParam()*, *setParam()*, *reset()*, *getState()*, and *getName()*. The *Concrete-Sensor* class represents a concrete sensor implementation of a measurement module. In our case, there are classes DS18x20, Dht22, Pt1000, Adc24bit, and others. These classes implement all the other methods defined in the *SlaveSensor* interface. These are mainly the *getValue()*.

3.2 Server Modules

Server modules form a vital part of the sensor IS, as they ensure that the measurement service is reliably started on the modules and that the readings are taken and stored in the database model. The second important part is enabling uniform access to data and measurement service settings.

3.2.1 nSoric API

The system's data model is divided into two parts: the administration of installations, users, and their permissions (Fig. 3.1) and the model of the measurement system instance itself (Fig. 3.2). There is just one model of users and settings in the whole sensor system. It contains records for all the users (from different installations) and their assigned applications or instances of measurement systems. Information about all installations of the sensor system is stored in this model. Each sensor system installation has its assigned database (Fig. 3.2), where that instance's measured values and configuration are stored.

Only the JSON format is used for communication. This format allows easy data transformation from the database tables into object notation. The length of the API server response depends on the data being sent. When sending information from configuration tables, the length of the message is usually up to 1kB. When a request is made to send the measured data, the message size can increase to the order of hundreds of kB up to units of MB, which is not an insignificant length and can form a system bottleneck on a slow connection. Therefore, all the server responses will be compressed using zip compression. For correct transmission over the text-oriented protocols, this message will be base64 encoded.

The nSoric API server service is designed as a REST API that contains four separate parts (Figs. 3.5 and 3.6):

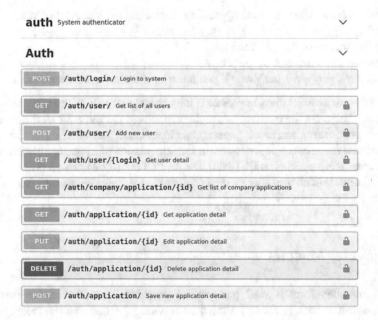

Fig. 3.5 Design of the REST API. Authorization for the nSoric API services

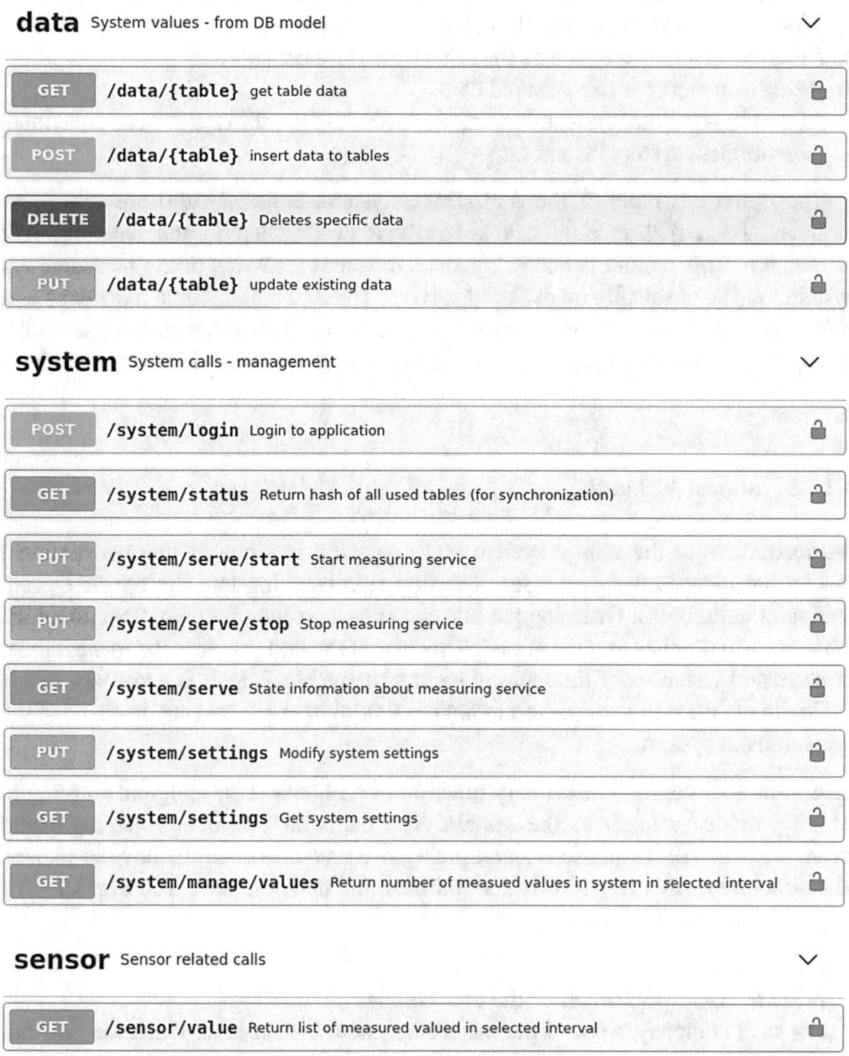

data System values - from DB model   ⌄

| GET | /data/{table} get table data | 🔓 |

| POST | /data/{table} insert data to tables | 🔓 |

| DELETE | /data/{table} Deletes specific data | 🔓 |

| PUT | /data/{table} update existing data | 🔓 |

system System calls - management   ⌄

| POST | /system/login Login to application | 🔓 |

| GET | /system/status Return hash of all used tables (for synchronization) | 🔓 |

| PUT | /system/serve/start Start measuring service | 🔓 |

| PUT | /system/serve/stop Stop measuring service | 🔓 |

| GET | /system/serve State information about measuring service | 🔓 |

| PUT | /system/settings Modify system settings | 🔓 |

| GET | /system/settings Get system settings | 🔓 |

| GET | /system/manage/values Return number of measued values in system in selected interval | 🔓 |

sensor Sensor related calls   ⌄

| GET | /sensor/value Return list of measured valued in selected interval | 🔓 |

Fig. 3.6 Design for nSoric API services (data, system, sensor)

1. API.data—access to data from the database. It is used to synchronize the local copy of the database, as well as to insert and edit data.
2. API.system—management of the nSoric measurement server.
3. API.sensor—access to measured data,
4. API.auth—user authentication controls access to the entire system and individual measurement system instances.

The individual parts of the API (Data, System, Sensor, Auth) are created as modules. Each API module can be enabled or disabled in the resulting API installation. This feature is necessary since different instances of the measurement system may be physically on different servers. The API.auth module must exist as a single instance since it provides central access to the other instances. For the other instances of the sensor system, the API.system, API.sensor, and API.data modules are required.

3.2.1.1 nSoric API.auth

Authorizations to the nSoric system and a specific instance of the measurement system are independent processes. The first step is to log into the nSoric system (endpoint auth/login). Once logged into the system, a list of applications to which the logged-in user has granted access. This part of the API provides the management of users and instances of the measurement system (Fig. 3.5). It is necessary to use the log in via the API.system—endpoint system/login to log into any instance of the measurement system.

Endpoint auth/login The primary function of API.auth is the endpoint auth/login, which provides a login to the system. The endpoint parameters are login and password. The API response is a comprehensive response about the logged-in user. The structure of this response is derived from the database design in Fig. 3.1. The response contains the following:

- User details
- A list of companies to which the user belongs
- For each company, a list of the metering system instances to which the user has access
- An authentication token for logging into the target instance of the measurement system

In the Listing 3.1, there is a sample endpoint response auth/login.

```
{
  "user": {
    "login": "juraj.dudak",
    "name": "Juraj Dudak",
    "isAdmin": 1,
    "companies": [
      {
        "id": 10,
        "name": "MyCompany",
        "enabled": 1,
        "jwtSecret1": "__secret1__",
        "applications": [
          {
            "name": "home measurement",
            "host": "http://place7.nsoric.com",
            "dbHost": "localhost",
            "dbUser": "server_user",
            "dbPassword": "SECRET_DB_PASSWORD",
            "application": "instance3.nsoric.com",
            "jwtSecret2": "__secret2__",
            "server": {
              "name": "saturn",
              "apiSchema": "https",
              "apiPort": 81,
              "endpoint": "api2"
            }
          }
        ]
      }
    ]
  },
  "JWT": "JIOIJH.ADSPIDUSAP.SJDOASID"
}
```

Listing 3.1 API response for auth/login authorization request

Endpoint auth/login forms an integral part of logging into a particular instance of the measurement system. Part of the response includes information about the URL of the specific instance of the measurement system (applications [i] .application), the server on which the instance is running (applications [i] .host), the database information of the specific instance (applications [i] .dbHost). Other API parts need this data to connect to the specified instance. The response further contains a JWT attribute, which is an authentication token. "Bearer Token Scheme" type authentication is used when communicating with other endpoints. The JWT token must be specified in the header. The token contains the information: *login*, *IsAdmin* flag, and *AppKey*, which is a part of the 2-step authorization for a specific instance of the measurement system. In addition to the login functionality that the API provides, additional functionality is available for managing other entities—adding, editing, and deleting:

- User management (auth/user)
- Management of measurement system instances (auth/applicationID)
- Company management (auth/company)

The process of authentication to the other parts is described in Sect. 3.2.1.3.

3.2.1.2 nSoric API.data

Endpoints /data/{table} (Fig. 3.6) are created, where {table} represents the table name (see Fig. 3.2) to access the data in the database. The given endpoints can be applied to all the tables except the sensor values, which represent the measured values. Meaning of each endpoint:

- [GET] /data/{table}—returns the contents of the requested table in the JSON format. The structure of the response is different for each table. The endpoint has no additional parameters.
- [POST] /data/{table}—Endpoint requires a mandatory part: the record to be added. This record is in the request body in JSON format.
- [DELETE] /data/{table}—Endpoint requires one URL parameter "id"—the value of the primary key of the record to be deleted.
- [PUT] /data/{table}—Endpoint requires one URL parameter "id" (ID of the record to be modified) and an object in the JSON format (new properties of the modified record) in the request body.

A sample of the API response for the [GET] /data/sensor endpoint is in the Listing 3.2. This is to retrieve the contents of the *Sensors* table. The response contains a context section, which contains the name of the source table, and a data section, which contains the actual rows of the requested table in a simple list format. The structure of this list is different for each table.

```
{
  context: Sensors,
  data:
  [
    [uid1, name1, level, minValue, maxValue, typeId, devideId],
    [uid2, name2, level, minValue, maxValue, typeId, devideId],
    ...
  ]
}
```

Listing 3.2 Response format for endpoint /data/sensor

3.2.1.3 nSoric API.system

The second part of the API functionality concerns the management of the nSoric serve measurement service. It is possible to check and change the status of the service, edit settings and get the information needed to synchronize the database using the API calls. The meaning of each endpoint is the following:

- [GET] `system/status`—each table in the database returns the calculated hash. This hash is used when comparing the contents to the local copy of the database. The end application can then request only the contents of those modified tables.
- [GET] `system/settings`—returns the object with the current setting. This is the setting of the measuring modules: their addresses, the method, and the parameters of connection to the measuring PC.
- [PUT] `system/settings`—modification of measurement system instance settings.
- [PUT] `system/serve/start`—launch of the nSoric serve measurement service.
- [PUT] `system/serve/stop`—stopping of the nSoric serve measurement service.
- [GET] `system/serve`—Information about the status of the measurement service (started/stopped). A sample response format is in the Listing 3.3.

```
{
    context: system,
    data:
    {
        serve: 1
    }
}
```

Listing 3.3 Response for auth/login authorization request

Authentication A 2-stage algorithm, similar to OAUTH-2, has been proposed for logging into the measurement system instance. The login process is as follows:

1. The access point for login is always API.auth (auth/login). The API server response is information about the applications the user can log into. The response includes a JWT token to be used for further communication.
2. As a part of the 2-step authentication, the client logs into the endpoint to log into the API.data (system/login) section. The URL of this endpoint is found in the response in step 1. This request must include the identifier of the target application, the URL of the target application, and the JWT token from step 1 that is used for authentication. The login name is a part of the JWT token.
3. The response consists of structured information about the authenticated user. A sample response is in the Listing 3.4. It contains the parts:

   a. Information on the authenticated user
   b. A list of parts to which the user has access—"areas," for each area, a group of sensors called "sectors" is defined
   c. For each sector, there is a list of available sensors

d. The JWT token is necessary for further communication with API functions in API.data, API.sensor and API.system. It is marked as *JWT2* in the response

```json
{
  "user": {
    "login": "juraj.dudak",
    "name": "Juraj Dudak",
    "group": {
      "name": "maintainer",
      "allowView": 1,
      "allowManage": 1,
      "allowAdministration": 0
    },
    "areas": [
      {
        "name": "area 51",
        "sectors": [
          {
            "name": "outddor sector",
            "sensors": [
              {
                "id": "__sensor_db_id__",
                "name": "humidity",
                "uid": "12adc021980cbfe9",
                "min_value": 0,
                "max_value": 100
              }
            ]
          }
        ]
      }
    ]
  },
  "JWT2": "JIOIJH.ADSPIDUSAP.SJDOASID"
}
```

Listing 3.4 API response for the authorization request

3.2.1.4   nSoric API.sensor

In the third part of the API, there is a single endpoint that ensures that the measured data is sent at the selected interval. Endpoint /sensor/value has two query parameters of the *datetime* type (*from*, *to*), which specify the time interval when selecting the measured data from the database, and an optional parameter sensor, which contains the ID of the sensors for which the data should be sent. If the *sensors* parameter is not specified, the response will contain data from all the existing sensors in the database. The format of the response is in the Listing 3.5. The response contains a list of sensor IDs for which the data was requested— the context section. The data section contains the measured values in the format *datetime:valueObject*, where *datetime* represents the measurement time stored in the *MeasuredDatas* table. The *valueObject* object represents a list of measured values from the requested sensors for that measurement time. The expression *sen* represents a single record of the *MeasuredValues* table (id, sensorUid, dateId, value)—see Fig. 3.2. In the case of the Listing 3.5, the measurements on sensors sensorId1, sensorId2, sensorId3 at two time intervals, *datetime1* and *datetime2*, are displayed. In each time interval, measurements were taken on all three sensors.

```
{
  context:[sensorId1, sensorId2, sensorId3],
  data:
  {
    datetime1: [
      [sen.id1, sen.sensorUid1, sen.dateId1, sen.value],
      [sen.id2, sen.sensorUid2, sen.dateId1, sen.value],
      [sen.id3, sen.sensorUid3, sen.dateId1, sen.value],
    ],
    datetime2: [
      [sen.id4, sen.sensorUid1, sen.dateId2, sen.value],
      [sen.id5, sen.sensorUid2, sen.dateId2, sen.value],
      [sen.id6, sen.sensorUid3, sen.dateId2, sen.value],
    ],
    ...
  }
}
```

Listing 3.5 Response format for endpoint /sensor/value

3.2.2 nSoric Serve

In automatic mode, when the continuous readings from the measurement modules are required, it is necessary to ensure that the operating system service level software, which will provide this measurement, is running. The proposed software has been named nSoric serve. The requirements for this type of software can be summarized as the following points:

- In the event of an unexpected error, the software must be capable of the following measurement.
- The software must be executable in the CLI mode (command line interface).
- The software must communicate with connected hardware blocks with constraints that are defined in the database.

The implementation used the Java programming language, which is known for its strict handling of events that can cause an error. This mechanism is called exception handling. An example of a potentially dangerous operation is an attempt to connect to the measurement module. In the case that a communication error, connection error, or other unexpected error occurs, actions must be defined that

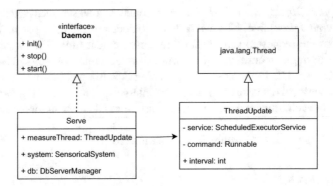

Fig. 3.7 Class diagram for the nSoric serve solution

can handle this condition. The requirement for continuous operation is met by the design of the solution. In Fig. 3.7 is shown the class diagram of the solution: the resulting and executable class is Serve. This class implements the methods defined in the *Deamon* interface from the *org.apache.commons.daemon* package. Apache Commons Daemon software is a set of utilities and Java support classes for running Java applications as server processes. Those are commonly known as "daemon" processes in Unix terminology (hence the name). On Windows, they are called "services." Daemon provides a portable means of starting and stopping a Java Virtual Machine (JVM) that is running server-side applications. Such applications often have additional requirements compared to client-side applications.

The measurement is run as a separate computational thread—the *ThreadUpdate* class, which is a descendant of the Thread class. The *ThreadUpdate* class ensures that the measurement process is run repeatedly at a time interval defined by the interval property. The measurement process itself is defined in the *Serve* class. The principle of operation of the sNoric Serve is the following:

1. Connecting to the database. Selecting the information about the measurement modules and sensors to be addressed.
2. For each measuring module:

   a. Connecting to the measuring module. Each module may have a different connection method.
   b. Starting the measurement on the sensors of the measuring module.
   c. Closing the connection to the measuring module.

3. Saving the measured data to the database.
4. Close the database connection.
5. Waiting for a defined period (*ThreadUpdate*, property interval); start from step 1.

As mentioned, the nSoric Serve software has been programmed in Java, which guarantees cross-platform use. Thus, this program can be run on GNU/Linux OS, as well as in Windows. The Java Runtime Environment is installed as a prerequisite. The resulting program serve.jar has, in addition to the preferred way of using it as a system service, the possibility to run it from the command line (Listing 3.6).

```
java -jar Serve.jar D <server> <user> <pass> <db>
java -jar Serve.jar I <server> <user> <pass> <db>
java -jar Serve.jar H <port> <address>
```

Listing 3.6 Using Serve on the command line

In the command line mode, Serve has three modes: D (database), I (Information), and H (Hardware).

Mode D(atabase)—the `<server>`, `<user>`, `<pass>`, and `<db>` parameters represent the database server address, login name, password, and database name, respectively. When started, the Serve software will only address those measurement modules and sensors stored in the database.

Mode I(nformation)—the meaning of the parameters is the same as in the D mode. In this case, the connected hardware (measurement modules) is not addressed, but the last measured values are read from the database.

Mode H(ardware)—in this mode, the software attempts a measurement on the measurement module with the specified address and communication port. In this case, the database entries are ignored. A single measurement is taken on all the sensors in the connected measurement module.

3.3   Software Applications

To work with the hardware modules, or measurement modules, a service software has been designed and programmed, which is used to configure these modules, direct reading of values, display, and partial analysis of existing data. In designing the new applications, common requirements were defined, which must be fulfilled by each piece of software:

• The application must be platform-independent.
• The application must have a simple user interface.
• It must use the common nSoric senlib library for data access.
• It must have an automatic update mechanism.

All the applications presented here have been programmed in Java (version JDK8) and the Swing framework. When communicating with the measurement modules or with the nSoric API, the established nSoric senlib library is used. A repository with available releases of the application versions has been created to ensure the implementation of the automatic update requirement. When the application is started, a request is sent to this repository (or its API interface) for information about the last available version of the application. If there is a newer version in the repository than the running application, the application will request to download the newer version and starts the update process. When the application restarts, the updated version is installed.

3.3.1 nSoric Merula

The nSoric Merula software work (Fig. 3.8) is a desktop application for direct measurements on the measurement modules of a sensor system. It supports connectivity via the USB (virtual COM port) or the IP address. It can communicate with multiple measurement modules. The primary purpose of this application is to trigger on-demand measurements on supported measurement modules.

Supported Measurement Modules The nSoric Merula software uses the nSoric senlib library, which implements the uBUS application protocol, to communicate with the measurement modules. The connection of the measurement modules is implemented via a USB/RS-485 converter. The connection parameters are baud rate = 19,200 baud, parity = 1, 8 data bits, 1 stop bit. Each measuring module has a unique address (in the local sensor network). Valid addresses are from 0x0010 to 0xFFF0. The module address is configurable and can be modified in the nSoric Cofig software. The following measurement modules are currently supported:

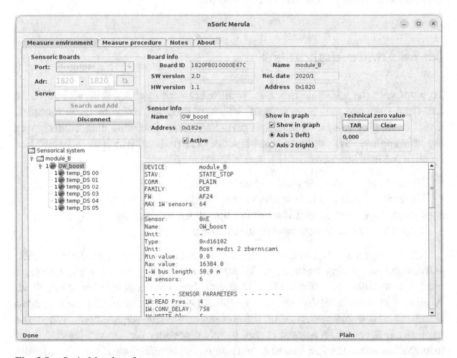

Fig. 3.8 nSoric Merula software

- OWB—measuring module with improved 1-wire bus controller. This module supports 1-wire bus lengths up to 600 m. In synergy with Merula, it can detect the 1-wire bus length and correctly set the 1-wire bus timing parameters to achieve reliable communication. It supports up to 128 sensors on the 1-wire bus.
- RRM—receiving module for the RTM wireless sensors. The module uses an operating frequency of 433.92 MHz. It supports up to 64 RTM sensors.
- TNZ—module for strain gauge measurements. The module includes two configurable channels for the strain gauge or temperature measurements with the PT1000 sensors. The measurement frequency is configurable: 10 Hz or 16.6 Hz.
- THB—module for measuring the relative humidity and temperature on up to four independent sensors.

3.3.1.1 Properties of the nSoric Merula Software

Connection Mode nSoric Merula supports local and remote communication. On GNU/Linux, automatic discovery of local communication devices is available. Supported communication interfaces:

- Local connection via virtual COM port (/dev/ttyUSBx in Linux, or COMx in Windows)—Fig. 3.9a
- Remote communication. In the case of using an ETH/Serial converter, it is possible to connect through the IP address and port—Fig. 3.9b
- Demo connection. It uses the local loopback to showcase software functionality without connection to the actual hardware.

In connecting to the measurement modules, a range of addresses is specified within which the application tries to find the connected measurement modules. The application connects to all the measurement modules whose address is within the searched range.

Once connected, information about the connected measurement modules is displayed: In the left-hand part of the application window (Fig. 3.8) there is a tree

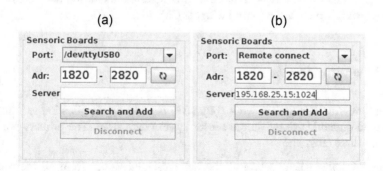

Fig. 3.9 nSoric Merula—connection modes

view of the hierarchical structure: a list of measurement modules with connected sensors or additional interfaces. In Fig. 3.8, the measurement module "Module_B" is shown, which contains the interface "OW_boost," which represents the 1-wire bus. Four sensors are detected on this bus. After clicking on the measurement module, the "Board info" section displays the selected measurement module information: module ID, SW/HW version, module designation, module name, date of manufacture, and module address. Similarly, if a sensor is selected in the sensor system tree, its properties are displayed in the "Sensor info" section: the sensor name and address. Selecting "Active" causes the sensor to be included in the measurement. When displaying the measured data in a live graph, it is possible to choose whether the measured value will be displayed and, if so, which axis will be used. At the bottom of the window is an area with informative statements during the initialization of the communication and the actual measurement on the measuring module.

3.3.1.2   Measurement Procedure

The measurement can be started from the second tab, "Measure procedure." The basic settings for the measurement are:

- Enable/disable live graph from measurement
- Delay between measurements in seconds
- Graph size: the maximum number of data points displayed in the graph
- Output format: CSV (comma-separated values) or UDF (contains CSV part and user notes related to measurements)

When the measurement is started, the data from the sensors (Fig. 3.10), marked as active, will be displayed. A legend is displayed below the graph. Sensor names are taken from the configuration of the measurement module or its sensors. In Fig. 3.10 is shown the measurement waveform on the OWB module—i.e., on the sensors of the 1-wire bus. The status bar shows information about the 1-wire bus itself: the operating voltage $U(ow)$ level and the current flowing through the bus $I(ow)$. The last entry is the measurement serial number.

After the measurement is completed, it is possible to add the text notes to the measurement, which will be saved with the CSV file.

3.3.2   nSoric Aurela

The nSoric Aurela software work (Fig. 3.11) is a desktop application to visualize measured quantities stored in the sensor system data store. The primary window contains three tabs:

- Current state—visualization of the last measured values from all the sensors in the sensor system. The following data is available for each sensor: unique

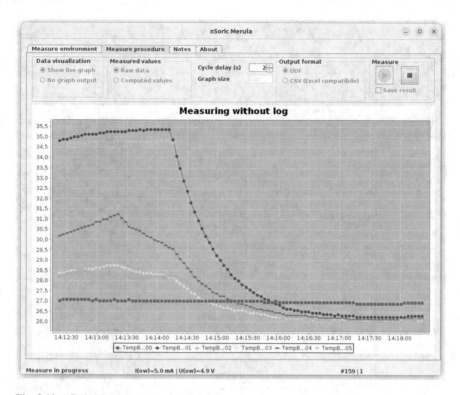

Fig. 3.10 nSoric Merula—measuring of data

sensor address, sensor location, time and value of the last measurement, graphical representation of the value for better readability, and, for battery-powered sensors, a visualization of the battery status.
- Time records—displaying sensor measurement graphs according to the sensors' division into areas and sectors (Fig. 3.12).
- System log—recording of all the interventions in the sensor system: adding a measuring module/sensor, changing the parameters of a module or a sensor.

In Fig. 3.12 are shown the time histories from the sensor system installation at the University of Žilina (Slovakia), where 3 OWB measurement modules and 32 temperature sensors are installed in horizontal and vertical mountings. The horizontally mounted groups are installed at the same depth in the body of the asphalt pavement. The vertical groups are located in the same geographical position but with different depths of embedment.

An essential part of the nSoric Aurela software is the administration of the measurement system instance, where all the parameters and system settings can be changed. There is the management of users and their permissions, management of the measurement modules and sensors, division of sensors into groups, and administration of the nSoric service itself.

Fig. 3.11 The nSoric Aurela software

3.3.2.1 Administration of Sensor System Settings

The structure and hierarchical relationships between different entities of the sensory system are described in Sect. 3.1.1. In the administration part of the nSoric Aurela software, it is possible to edit the following settings.

Users—Creating and editing users in the system. Allows modifying the permissions to view the data, modifying parameters of measurement modules, and access to the administration.

Areas—Sensors are categorized into a 2-level hierarchy: areas and sectors. An area represents a geographical location where sensors can be divided into smaller clusters, referred to as sectors. Within a single region, there can be one or more

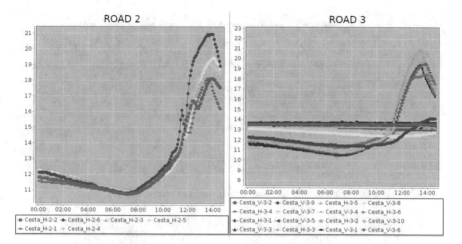

Fig. 3.12 The nSoric Aurela software—recording of the measured data

Name	Type/Address	Sector	Min	Max	Alarm Lo	Alarm Hi
Cesta_H-2-2	digitalny_teplomer_ds18b20	Cesta 2	-20.0	80.0	-0.5	40.0
Cesta_H-2-6	digitalny_teplomer_ds18b20	Cesta 2	-20.0	80.0	-0.5	40.0
Cesta_H-2-3	digitalny_teplomer_ds18b20	Cesta 2	-20.0	80.0	-0.5	40.0
Cesta_H-2-5	digitalny_teplomer_ds18b20	Cesta 2	-20.0	80.0	-0.5	40.0
Cesta_H-2-1	digitalny_teplomer_ds18b20	Cesta 2	-20.0	80.0	-0.5	40.0
Cesta_H-2-4	digitalny_teplomer_ds18b20	Cesta 2	-20.0	80.0	-0.5	40.0
Cesta_V-3-2	digitalny_teplomer_ds18b20	Cesta 3	-20.0	80.0		
Cesta_V-3-9	digitalny_teplomer_ds18b20	Cesta 3	-20.0	80.0		
Cesta_H-3-5	digitalny_teplomer_ds18b20	Cesta 3	-20.0	80.0		
Cesta_V-3-8	digitalny_teplomer_ds18b20	Cesta 3	-20.0	80.0		
Cesta_H-3-4	digitalny_teplomer_ds18b20	Cesta 3	-20.0	80.0		
Cesta_V-3-7	digitalny_teplomer_ds18b20	Cesta 3	-20.0	80.0		
Cesta_V-3-4	digitalny_teplomer_ds18b20	Cesta 3	-20.0	80.0		
Cesta_H-3-6	digitalny_teplomer_ds18b20	Cesta 3	-20.0	80.0		
Cesta_H-3-1	digitalny_teplomer_ds18b20	Cesta 3	-20.0	80.0		
Cesta_V-3-5	digitalny_teplomer_ds18b20	Cesta 3	-20.0	80.0		
Cesta_H-3-2	digitalny_teplomer_ds18b20	Cesta 3	-20.0	80.0		
Cesta_V-3-10	digitalny_teplomer_ds18b20	Cesta 3	-20.0	80.0		
Cesta_V-3-3	digitalny_teplomer_ds18b20	Cesta 3	-20.0	80.0		
Cesta_H-3-3	digitalny_teplomer_ds18b20	Cesta 3	-20.0	80.0		

Fig. 3.13 The nSoric Aurela software—sensor administration

areas. An area represents a logical grouping of multiple sensors, either by actual location or the placement method (horizontal or vertical).

Sensors—List of registered sensors in the sensory system. For each sensor, it is possible to define the designation or a name, location in the sector, minimum and maximum measured values, and the values at which the alarm is activated for the given sensors. The sensors belonging to the "testing01" area are shown in Fig. 3.13. In the right-hand part, there is an option to assign a sensor to an existing sector. The alarm values are used when displaying the instantaneous values—Fig. 3.11— the last value column. In addition, if the value of a sensor exceeds the alarm value, information about the sensor that caused the alarm is displayed at the top of the window.

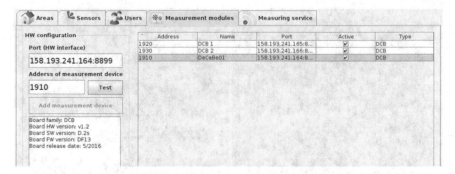

Fig. 3.14 The nSoric Aurela software—measuring modules administration

Fig. 3.15 The nSoric Aurela software—measurement service as administration

Measurement Modules—This section presents a list of the registered sensor modules. The record for each sensor module contains information about its address, name of the connection method (*Port*), and type—Fig. 3.14. Measurement modules can be deactivated—in practice, this means omitting the measurement module from the measurement process. Therefore, that is a software deactivation. A new measurement module can only be added to the system by verifying its existence and functionality. In the prepared editing fields, one needs to enter the connection method (*Port*)—in this case, that is the IP address of the Ethernet/serial converter to which the measurement module is connected. The second mandatory entry is the address of the measurement module. After entering these data, the "Test" function is available, which verifies the module's existence and retrieves the basic information about the connected module and a list of sensors that this module contains. The measuring module and its sensors can then be added to the sensor system.

Measuring Service—The last part of the administration is the management of the nSoric serve measurement service, which runs on a remote server (Fig. 3.15). When the administration window is opened, a question is sent to the nSoric API asking about the status of the measurement service. The result is then displayed. The measurement service can be stopped or started remotely.

3.3.2.2   Data Synchronization

The principle of the application functioning is client/server. All the valid data is always on the server. When the application is started, a local copy of the data and settings is created. When an inconsistent state is detected, such as data on the server differ from the data in the local copy, the data on the server is always considered valid. The data in the application is divided into two groups, which differ in the way of synchronization:

- Configuration includes data from all the tables (Fig. 3.2) except measured data. The data from these tables are synchronized each time the application is started.
- Data—data is synchronized on demand. When displaying the data, the user can select the time interval for display (Fig. 3.11, top left—select box "time interval").

When the settings in the administration part are changed, it proceeds to save the change made to the server. We show the implemented algorithm for reliable editing of records:

1. After the configuration change is made, a request is sent to the nSoric API, the communication interface with the sensor system instance.
2. The request is executed on the server. The response to the request is the result of the performed action.
3. According to the result of the request, the values are written/changed in the local settings.

3.3.3   nSoric Cofig

The nSoric Cofig software (Fig. 3.16) is a desktop application designed to configure measuring modules. It can retrieve the configuration stored in the measurement module, save it as a data file, modify an existing setting or create the new one and save this setting to the measurement module. Additional functions include testing the correct functionality of the measurement module and creating a pdf log of the module configuration.

Each measurement module has 1kB of the microcontroller's persistent memory reserved for its configuration. The organization of this memory block is described in Sect. 1.3.2. The primary division of the configuration memory is shown in Fig. 1.19. The primary task of the nSoric Cofig software is to work with the configuration of the measurement module. The main application window (Fig. 3.16) is divided into several parts:

- Connection—setting the parameters of the connection to the measuring module. These settings are the same as in the nSoric Merula software.
- Measuring board—contains actions for connecting/disconnecting the measurement module, reading the configuration from the measurement module, and

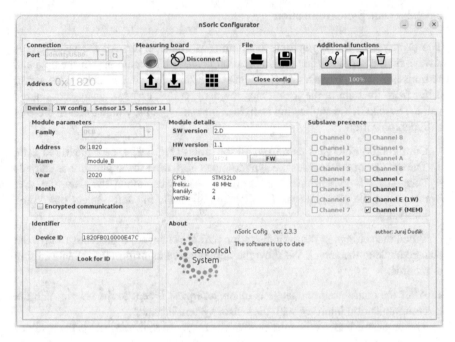

Fig. 3.16 The nSoric Cofig software

uploading the configuration to the module. The last button makes the contents of the configuration memory available as the unprocessed ("raw") values.

- File—Loading and saving the configuration as a file. The configuration file is an exact binary copy of the settings, as stored in the measurement module.
- Other functions—diagnostics of the measuring module, creating a pdf log of the measuring module settings, and resetting all the settings. For the diagnostics function, a set of standard requests is prepared to which the measuring module must respond.
- Device—information about the measuring module. This information is identical to the information in the nSoric Merula software. The difference is that the module parameters can be edited using this software.

A measurement module can contain up to 16 sub-slave modules, designated as a channel in the nSoric Cofig software. In the "Sub-slave presence" section, it is possible to indicate which sub-slave modules are implemented in a particular measurement module. A special case of a submodule is the 1-wire bus communication interface (Fig. 3.17).

The 1-wire interface configuration is available for the OWB measurement module. Part of verifying the 1-wire interface configuration is to test the communication and scan the 1-wire bus. The "1W sensor list" section lists the detected 1-wire sensors connected to the measurement module. This sensor list can be exported to a CSV file. The nSoric Cofig configuration software supports communication

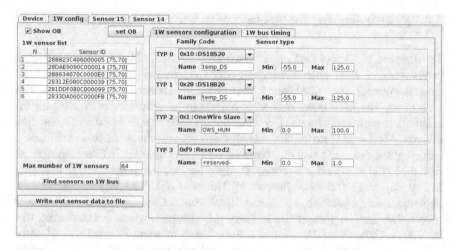

Fig. 3.17 The nSoric Cofig software—The 1-wire configuration

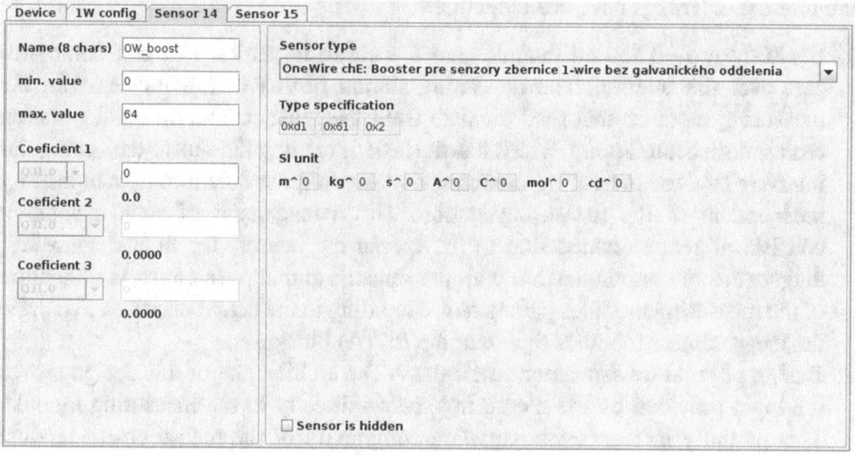

Fig. 3.18 The nSoric Cofig software—the sub-slave configuration

with the OWS sensor modules described in Sect. 2.1. In the middle part of the window is a list of supported 1-wire sensor types. In this case, the OWB module will support the DS18B20 and DS18S20 sensors—which are digital thermometers—and the OWS sensor. A range of valid measured values can be defined for the sensor type. The configuration of the standard sub-slave module is shown in Fig. 3.18. For each sensor, it is possible to set the specific type, the name of the sensor, the valid range of measured values, and the calibration coefficient that can be used in the conversion of the resulting value.

3.4   Sensor System Integration with the IoT Networks

The Internet of Things (IoT) is a network of physical objects-devices, vehicles, machines, and other objects with embedded electronics, software, sensors, and network connectivity—that enables these objects to collect and exchange data with each other. The IoT allows connected objects to be controlled remotely over the existing network infrastructures, creating opportunities for further direct integration of the physical world into the computer systems. The results are increased efficiency, accuracy, and economic benefits. There are significant trends in IoT development. At first, the IoT is presented as a service. On the other side stands the world of industrial IoT deployment—Industrial Internet of Things (IIoT). The IoT, as a service, is implemented as a solution, e.g., Smart Home, Smart City, Intelligent building, and others. The IIoT is part of the Industry 4.0 standard, where it is assumed that humans, computers, and machines can exchange relevant information and make better decisions based on this information. In the design, at each stage of the sensor system implementation, the three perspectives on the possibility of integration into the Internet of Things have been identified:

1. **Basic design**—The end modules are a measuring station that can communicate over the Internet. The measuring station further communicates with the measuring modules that read the data from the sensors. The measuring station represents a central point, which has the task of reading the sensor data at regular intervals (via the measuring modules) and sending the data to a data repository independent of the measuring station. The management of such a solution consists of remote connection to the measuring boards, for their diagnostics, diagnostics of communication with the sensors, and adjustment or modification of the measuring modules parameters, according to the actual needs. In Fig. 3.19, the measurement modules represent the IoT(A) blocks.
2. **Design of local measurement modules**—The architecture of this design moves the logic provided by the measuring station directly to the measuring module. This partially increases the hardware complexity of the measurement module solution but eliminates the intermediate measurement station. Each measurement module must be able to send its measured data to the data store, no longer worrying about whether the data has been stored. The remote management of the measurement modules is provided in a similar way as in the first case. A communication subsystem capable of sending data to the Internet shall be an integral part of the measurement module. In Fig. 3.19, the measurement modules represent the IoT(B) blocks. The adjective "local" means that such a module connects to the local network via Wi-Fi or a fixed Ethernet connection. The availability of the local network limits the location of the measurement module.
3. **Design of autonomous measuring modules**—The architecture is similar to a local measurement module. The difference is in the connection method. This module is not dependent on the local network. It contains a communication interface allowing it to communicate directly with a remote server, using other

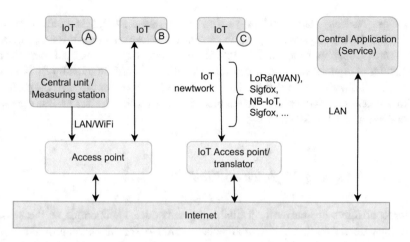

Fig. 3.19 Communication in IoT networks

technologies, such as Wi-Fi or Ethernet. These are LoRa/LoRaWAN, Sigfox, Wireless M-Bus or NB-IoT technologies. In Fig. 3.19 such a module represents IoT(C).

Here is a brief description of IoT communication technologies:

LoRaWAN—The LoRaWAN®specification is a Low-Power, Wide-Area (LPWA) networking protocol, designed to wirelessly connect battery operated "things" to the Internet in regional, national, or global networks. It targets the critical Internet of Things (IoT) requirements, such as bi-directional communication, end-to-end security, mobility, and localization services. The LoRaWAN®network architecture is deployed in a star-of-stars topology in which gateways relay messages between end devices and a central network server. The gateways are connected to the network server via the standard IP connections and act as a transparent bridge, simply converting the RF packets to IP packets and vice versa.

NB-IoT (Narrowband IoT)—is a specialized cellular network for the Internet of Things. It is a specially adapted cellular network that can take advantage of unused (but licensed) bands in the radio frequency spectrum. Most carriers reserve a small frequency band from the LTE band, but the NB-IoT also enables them to use a guard band (a band between two others) or other unused bands. The Nb-IoT was designed to address some of the past challenges of using cellular connectivity in IoT devices, namely its power consumption.

Sigfox—Sigfox 0G technology harnesses the Power of Low to deliver Massive IoT Energy efficiency, long-range, small data packet, low cost, simplicity, and security. Sigfox is the world's leading connectivity provider for the Internet of Things (IoT). Sigfox can pick up tiny signals sent by IoT devices worldwide—using the lowest amount of energy to the point where the natural energy harvesting technology will be enough to power the data transmissions.

Wireless M-Bus—The wireless version of Wireless M-Bus is defined by the EN13757-4:2013 [1] standard, which specifies wireless data transmission between the sensors, such as water meters, gas meters, electricity meters, heat meters, and concentrators/gateways, with a central evaluation system. Practically, it is about replacing the wiring of the physical layer of the OSI model of the M-BUS protocol with wireless communication and adapting the link layer. Wireless M-Bus operates at 868 MHz and 169 MHz.

3.4.1   Measuring Modules and IoT

Microcontrollers are currently available in embedded applications or the sensor solutions described here. An architecture suitable for implementing the measurement modules and sensor solutions is, e.g., ARM Cortex-M. The Cortex-M processor family is based on the M-Profile Architecture that provides the low-latency and highly deterministic operation for deeply embedded systems. Table 3.1 provides characteristics of different variants of this architecture.

The advantage of solutions built on STM32 microcontrollers is excellent performance scalability—from solutions using the Cortex-M0 core, suitable for the low-power sensor modules, to solutions built on the Cortex-M7 core, suitable for demanding applications. In the category of microcontrollers built on Cortex-M, there are two development branches: the STM32Fxx—the main development branch, and the STM32Lxx—the branch aimed at minimizing power consumption. The fact is that most of the microcontrollers of the STM32Fxx and STM32Lxx

Table 3.1 Cortex-M architecture description

Architecture	Description
Cortex-M23	The Arm Cortex-M23 is the smallest and lowest power microcontroller with TrustZone security
Cortex-M7	The Arm Cortex-M7 processor is the highest-performance member of the energy-efficient Cortex-M processor family
Cortex-M4	The Arm Cortex-M4 processor implements a good blend of control and performance for mixed-signal devices
Cortex-M3	The Arm Cortex-M3 processor is the industry-leading 32-bit processor for highly deterministic real-time applications
Cortex-M1	The Arm Cortex-M1 processor targets FPGA devices. It is available to access instantly with the DesignStart FPGA
Cortex-M0+	The Arm Cortex-M0+ processor is the most energy-efficient Arm processor available
Cortex-M0	The Arm indexARM Cortex M0Cortex-M0 processor is the smallest Arm processor available

family lack the WI-FI or Bluetooth communication interfaces, which are necessary for the integration of the proposed solution into the IoT network. Currently, microcontrollers using the Tensilica®Xtensa®LX platform, such as the ESP32 microcontrollers, are a popular solution. This solution's advantage is the built-in Bluetooth and WI-FI communication interfaces.

All the measurement modules presented in Chap. 2 are built on the 32-bit microcontrollers of the STM32F0x and STM32L0x family. Those microcontrollers belong to the "Value Line" and general-purpose groups, respectively. They contain the basic peripherals needed for communication with external sensors and other systems (PC or other communication modules), the RAM size is from 4 kB to 32 kB, and the FLASH memory size (for a program) is from 16 kB to 256 kB. The core frequency can be configured from 1 MHz to 48 MHz. The STM32L0x family is characterized by having lower power requirements and is suitable for solutions where the power consumption of the solution needs to be minimized. In the case of the STM32 microcontroller family, only the selected microcontrollers have hardware support for the Ethernet interface. Integrating the solution into the network infrastructure for other microcontrollers of the STM32 family can be implemented using dedicated modules or new types of microcontrollers with integrated IoT peripherals. The ST's solutions are divided into Long Range, Short Range, and RF. For integrating the existing solutions into the IoT networks, "Long-range" solutions are of interest. Solutions in this category include:

- LoRaWAN—wireless communication technology developed to create the low-power, wide-area networks (LPWANs) required for the machine-to-machine (M2M) and Internet of Things (IoT) applications. Existing hardware solutions:

  - STM32WL MCU series[2]—the world's first LoRa®-enabled System-on-Chip. The STM32WL is fully open and supports multi-modulation, making it the ideal choice for LPWAN and IoT developments with outstanding ultra-low power consumption without compromising performance.
  - P-NUCLEO-LRWAN2[3] (P-NUCLEO-LRWAN3[4])—hardware module for 868, 915, and 923 MHz RF (433 and 470 MHz RF) bands, comes with a gateway and an end-node.
  - I-NUCLEO-LRWAN1—expansion board for the STM32 Nucleo featuring a low-cost USI®LPWAN module supporting the LoRa®technology.

- Sigfox—An inexpensive, long-distance, lower-power solution primed for multiple industrial and consumer applications. The available solution from ST:

  - S2-LP[5]—Ultra-low power, high-performance, sub-1 GHz transceiver.

---

[2] https://www.st.com/en/microcontrollers-microprocessors/stm32wl-series.html.

[3] https://www.st.com/en/evaluation-tools/p-nucleo-lrwan2.html.

[4] https://www.st.com/en/evaluation-tools/p-nucleo-lrwan3.html.

[5] https://www.st.com/en/wireless-connectivity/s2-lp.html.

- STM32WL MCU series. Sub-GHz Wireless Microcontrollers. Dual-core indexARM Cortex M0Arm Cortex-M4/M0+.

- Wireless M-BUS—is an open standard developed for exceedingly power-efficient smart metering and Advanced Metering Infrastructure (AMI) applications, and it is quickly spreading in Europe for automatic electricity, gas, water, and heat meter reading applications. The available solution from ST:

  - SPIRIT1[6]—is a very low-power RF transceiver for wireless RF applications in the sub-1 GHz band. It is designed to operate both in the license-free ISM and SRD frequency bands at 169, 315, 433, 868, and 915 MHz.
  - STM32WL MCU series.

- KNX-RF—An open standard for residential and commercial building automation, the KNX communication protocol offers both wired and wireless node topologies, ensuring high interoperability between devices and systems without needing a central gateway.

  - STSW-S2LP-KNX-DK[7] is an evaluation package based on the S2-LP high-performance ultra-low power RF transceiver and BlueNRG-2 very low-power Bluetooth Low Energy (BLE) system-on-chip. It is designed to evaluate KNX-RF communication in the 868 MHz license-free ISM band.

Migrating an existing application between the microcontroller families is easy, thanks to the standardized STM32Cube[8] library. This library provides the same API for accessing the MCU peripherals across all the families. The library for each MCU family includes:

- The hardware abstraction layer (HAL) enables portability between different STM32 devices via standardized API calls.
- Low-layer (LL) APIs, a light-weight, optimized, expert-oriented set of APIs designed for both performance and runtime efficiency.
- A collection of middleware components including RTOS, USB library, file system, TCP/IP stack, touch-sensing library or graphics library (depending on the STM32 series).
- RF stacks such as Bluetooth®LE 5.2, OpenThread, Zigbee 3.0, LoRaWAN® and Sigfox, specific to every STM32 wireless series.

---

[6] https://www.st.com/en/wireless-connectivity/spirit1.html.

[7] https://www.st.com/en/embedded-software/stsw-s2lp-knx-dk.html.

[8] https://www.st.com/en/embedded-software/stm32cube-mcu-mpu-packages.html.

3.4.2 *Implementation of a Local Measurement Module*

The development of the measurement modules, described in Chap. 2, was focused mainly on the module's functionality according to the requirements in the specification. The functionality to connect the measurement module to the Internet was implemented as an afterthought. The variant "Basic measurement module" (Fig. 3.19, IoT-A module) was chosen to connect the measurement module to the Internet. The features of such a solution are:

- The module works without any firmware modifications for the microcontroller.
- The module uses RS845 serial interface for communication.
- The module includes a serial-to-IP address converter.

Figure 3.20 shows the basic design of the measurement module with remote access. The individual blocks have the following meaning:

Measuring module—measurement module, as described in Chap. 2.

Serial/ETH—embedded network server that maps a serial port to an IP address. That is a module containing at least one serial interface and at least one network interface. The network interface can be Ethernet or WI-FI. We list several existing reliable solutions:

- Lantronix UDS2100[9]—contains two serial ports that can be configured as RS-232, RS-422, or RS-485. Configuration of these serial ports is possible via the web interface or a dedicated application. It supports setting all the parameters for the serial port (number of bits, parity, start/stop bit, and baud rate). The baud rate for the serial port can be up to 921 KBaud.

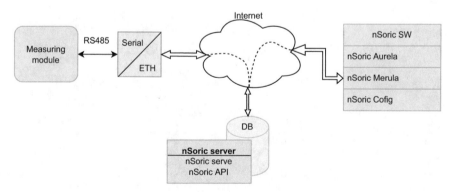

Fig. 3.20 Basic design of a measurement module with the remote access

---

[9] https://www.lantronix.com/products/uds2100.

- Lantronix xPico[10]—is the WI-FI module—a complete device server suitable for mobile M2M applications. The module features: WLAN interface IEEE 802.11 b/g and 802.11n (2.4GHz only), supports WPA-personal and WPA2-personal secure layer, includes two serial ports with a maximum data rate of 921.6 Kbps
- MOXA NPort P5150A[11] is a device server making simple and reliable serial-to-Ethernet solutions possible. It includes 1 Ethernet port and one serial port. The maximum speed of the serial port is 115,200 baud.
- MOXA NPort W2150A[12]—is a Wi-Fi module with support for 802.11a/b/g/n standards. It supports WEP, WPA, and WPA2 secure communication. The serial port can be configured as RS-232, RS-422, or RS-485. The maximum data rate is 921.6 Kbps.

nSoric server—a computer running the nSoric serve service to retrieve data from the measurement module and the nSoric API service to make the measured data available to end applications.

nSoric SW—a software application that communicates directly with the measurement module (nSoric Merula, nSoric Cofig) or requests the measured data from the server (nSoric Aurela). Since the services nSoric serve and API define a public software interface, it is possible to communicate with the measurement module from another application that uses this interface.

Reference

1. European Standard, CSN EN 13757-4, communication systems for meters - part 4: Wireless Mbus communication. CSN Standards EN 13757-4:2013, May 2019

---

[10] https://www.lantronix.com/products/xpico-wi-fi/.

[11] https://www.moxa.com/en/products/industrial-edge-connectivity/serial-device-servers/general-device-servers/nport-p5150a-series.

[12] https://www.moxa.com/en/products/industrial-edge-connectivity/serial-device-servers/wireless-device-servers/nport-w2150a-w2250a-series.

Chapter 4
Practical Use-Case of Proposed Measurement System

Practical installation is the best proof of concept.

The individual measurement modules were developed based on specific requirements from practice. This chapter gives examples of the real-life applications of the measurement modules within a sensor system. In all cases, the sensor data are collected and stored in a central data repository. Each installation includes a software part in which the actual measured values or time records from selected intervals are displayed. The nSoric serve software is used for the data acquisition, and the nSoric Aurela software is used for the data display and evaluation. We present 5 case studies:

1. Monitoring the effectiveness of interior insulation materials in heat transfer through the building wall
2. Monitoring the temperature in bulk biological materials
3. Monitoring the temperature in the production hall and visualizing the results in the form of a temperature map
4. Monitoring of the asphalt pavement and its subsoil freezing
5. Experimental measurement of fatigue of asphalt pavement

At the end of the chapter, five granted copyright utility models are listed, which were created in the design and implementation of individual hardware solutions. These solutions are registered at the Industrial Property Office of the Slovak Republic.[1]

1. Sensor for measuring mechanical deformation of asphalt pavement
2. Universal serial bus device for measuring physical quantities
3. Sensor for measuring the temperature profile of the asphalt pavement
4. System for wireless measurement of environmental variables in biologically active materials

---

[1] https://wbr.indprop.gov.sk/WebRegistre.

© The Author(s), under exclusive license to Springer Nature Switzerland AG 2023
J. Ďuďák, G. Gašpar, *Design and Implementation of Sensory Solutions for Industrial Environment*, Signals and Communication Technology,
https://doi.org/10.1007/978-3-031-30152-0_4

5. Rod probe for temperature measurement with an adjustable position of installed
   sensors

4.1   Monitoring of Ecological Building Insulation

Energy savings related to using new materials and technologies are now becoming
an inevitable standard in the construction industry. These trends are applied
in the construction of new buildings and the renovation of existing buildings.
From a thermo-technical point of view, the reduction of energy performance in
existing buildings is based on increasing the thermal resistance of the building
envelope and reducing the infiltration of windows and doors. Improvement of the
thermal insulation properties of external walls can be achieved through additional
insulation [1].

In most cases, the preference is to insulate the exterior of the building. The
composition of the wall is based on the placement of the lightweight thermal
insulating materials, which are well permeable to water vapor, on the outside of
the structure and heavier materials, which have the less thermal insulating capacity
and are less permeable to water vapor, on the inside of the structure. With this layer
composition, the building walls are protected against cold and moisture in the winter
months, condensation of water vapor occurs outside the structure, and the building
does not overheat in the summer months (Fig. 4.1).

However, in specific cases, it is not possible to use external insulation. It is most
often used in renovations of listed buildings and buildings where it is necessary
or desirable to preserve the original architectural appearance of the facade. This
was also the case in restoring the former people's school in Hrubý Šur,[2] Slovakia,
which was built in 1906. This building is planned to be used as EcoCentre, and

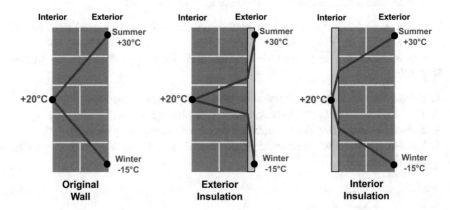

Fig. 4.1 Temperature course in the wall structure during winter and summer

---

[2] https://www.obnova.eu/portfolio/ekocentrum-artur-hruby-sur/.

ecological materials were used in the renovation. It is the first building in Slovakia to be insulated on the inside with insulation made of natural materials.

The interior insulation was designed only on the front facade using various natural materials such as straw, clay, sheep wool, and others. The rest of the building is insulated with straw bales. The temperature and humidity monitoring system, consisting of 104 digital thermometers and three hygrometers placed in the profile of the original wall and the different types of insulation, was designed to monitor the efficiency and functionality of the different types of thermal insulation in actual conditions.

4.1.1 Technical Solution

The installed system consists of basic measuring modules connected to the RS-485 serial bus, which can reach line lengths up to 1200 m. Each measuring module is a separate and independent unit with a unique address and processes data from the connected sensors. The OWB (Sect. 2.1) and THB (Sect. 2.3) measurement modules were used in this project (Fig. 4.2).

The nSoric system was installed in October 2014 and monitored the temperature for two years. The energy efficiency of the plaster types and insulation materials used is outlined in [2], where the paper's main aim was to present the technical solution of the sensor system. Measurement points were defined at which the temperature and humidity were to be measured to monitor the effectiveness of the building insulation with different types of plaster and insulation. The monitoring

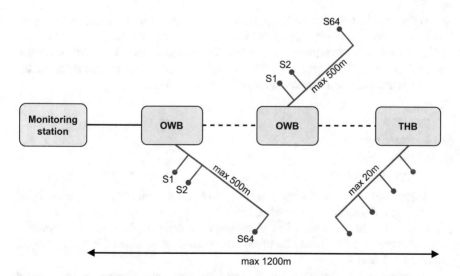

Fig. 4.2 Wiring diagram of the nSoric sensor system [1]

Fig. 4.3 Scheme of the wall
structure [2]

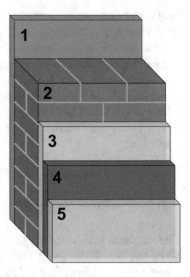

was carried out on the north wall of the building and in several layers. The layers
are defined as follows (Fig. 4.3):

- 1. Layer—external plaster
- 2. Layer—masonry
- 3. Layer—contact of masonry and insulation
- 4. Layer—contact of insulation and board
- 5. Layer—contact of board and plaster
- 6. Layer—internal plaster

Within the monitored building wall, ten essential measurement points were
defined. In Fig. 4.4, the measurement points are marked in red. At each measurement
point, six temperature sensors were placed. The insulated wall was divided into five
areas according to the type of insulation used. For the thermal insulation of the
masonry, the following insulation materials were used [2]:

- Wool
- Hemp
- Recycled textiles
- Crushed cork
- Blown fiber cellulose

There were two measuring points in each area, with each measuring point
containing 6 thermometers placed in individual layers.

In addition to the installation of temperature sensors, three relative humidity
sensors were installed in the building. The sensors were located in the 6th layer,
in areas 2, 3, and 4.

Based on the requirements for the monitoring process, a measurement interval
of every 10 min was chosen, i.e., 6 measurements per hour or 144 measurements

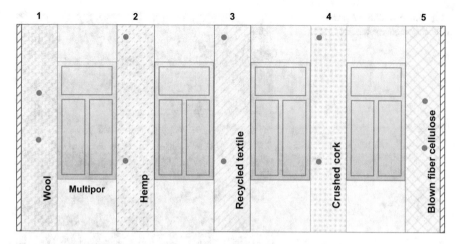

Fig. 4.4 Sensor placement in the different layers of the building wall [2]

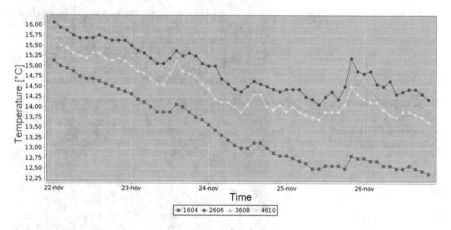

Fig. 4.5 Comparison of the insulation materials effectiveness

per 1 day. From each measurement, 104 readings are available. For one day, this is almost 15,000 measured data. When monitoring the temperature in the investigated building, the aim was to obtain the temperature profiles for different types of internal insulation.

To compare the effectiveness of the insulation materials used, the graph in Fig. 4.5 plots the temperatures from the 6th layer for different materials: 1604—wool, 2606—hemp, 3608—recycled textiles, and 4610—crushed cork. From this point of view, hemp comes out as the best traditional material for internal insulation. The measuring points were coded as follows: 1st digit—sensor location (according to Fig. 4.4), 2nd digit—sensor location in the layer within the measuring point, and 3rd and 4th digits—measuring point code.

Fig. 4.6 Installation of the measurement system

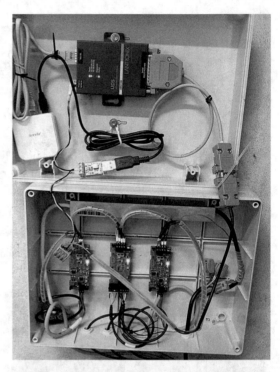

Fig. 4.7 Windows of truth—4th section with cork insulation and cork panel (before plastering) [3]

Figure 4.6 shows a photograph of a temporary, experimental installation of the measurement system. At the bottom are three measurement modules: 2x OWB module and 1x THB module (in the middle). The top part shows the communication part of the installation. There was no Wi-Fi network available in the area, so a 3G hotspot was used to ensure a stable connection. The Lantronix UDS1100 module is in the middle, providing the connection between the measuring modules (RS485 interface) and the network interface. This wiring ensured that the measurement modules were accessible via the Internet.

Figure 4.7 shows the cross-section of the insulated wall. Crushed cork insulation was used.

4.2 Temperature Monitoring in Bulk Biological Materials

Using biomass in the form of wood chips and wood pellets as an energy source is one of the most frequently used alternatives to the fossil sources used in heat production. Wood chips, bearing the characteristics of firewood, are produced from wood waste, branches, and raw materials obtained from thinning stands. Typical dimensions of wood chips range from 5 to 50 mm, 5 to 30 mm wide, and 5 to 15 mm thick. Wood chips are biologically active materials. They begin to decompose relatively quickly through the activity of living parenchymatous cells, chemical oxygenation, hydrolysis of cellulosic components in an acidic environment, and the biological activity of bacteria and fungi. This reduces the volume of wood chips produced and increases the moisture content of the material. At the same time, the temperature of the stored wood chips rises to between 50 and 70 °C and, under the right conditions, may even spontaneously ignite. The optimum moisture content of wood chips for combustion is in the range of 25 to 30%. If the water content of the chips is higher than the above range, the chips will degrade and mold after a specific time (depending on the temperature) [4].

The requirements for storing wood chips in the Slovak Republic are regulated by a decree.[3] It defines the need to measure the temperature of wood chips at a depth of 1.5 m at a distance no more than 10 m apart once a day. The measurement interval may be extended to once every 3 days if the temperature of the chips does not exceed 35 °C during the first week. After three weeks of storage, the measurement interval may be extended to once a week. If the temperature in a pile reaches 50 °C or increases by more than 3 °C in 24 h, the chips must be raked. In addition to the fire risk associated with the storage of wood chips, there are other risks, e.g., the occurrence of organic dust contaminated with microorganisms—bacterial and fungal spores.

From the above, it is evident that knowing the temperature of the stored wood chips is a rather demanding and responsible activity, especially in the case of large landfills. It is particularly problematic to monitor the temperature rise in the deeper layers of wood chips below the 1.5 m level set by the decree, which is accessible with a conventional stick thermometer since the height of the stored wood chips can exceed 5 m. Based on the requirements of landfill operators, we have therefore developed a system for monitoring the temperature field of woodchip landfills, consisting of radio frequency probes, receivers, and evaluation software [4].

---

[3] Decree of the Ministry of the Interior of the Slovak Republic No. 258/2007 on fire safety requirements for storage and handling of solid combustible substances.

Measurement System Requirements The primary goal in the design and manufacture of the measuring system was the complete automation of measuring and evaluating the temperature of the measured material, i.e., the total elimination of manually performed measurements in the biomass, employing the rod thermometers. The proposed system consists of measurement hardware, storing and archiving the measured data, and application user software. The individual parts of the measurement system have been designed to meet the following requirements defined by biomass processors:

- Temperature measurement from—$10\,°\mathrm{C}$ to $90\,°\mathrm{C}$ with an accuracy of $\pm\,0.5\,°\mathrm{C}$ and a resolution of $0.1\,°\mathrm{C}$
- Autonomous operation of the measurement probes for a minimum of three months
- Probe construction resistant to the pressure of the stored wood chips and mechanical stress from the mechanisms used in the processing and handling of the wood chips
- Operation of the probes in the free 433.92 MHz band
- Unidirectional communication protocol, allowing collision detection of duplicate transmissions of multiple probes on the receiver side
- Parameterized frequency of probe transmission depending on the temperature of the chip
- Probe dimensions at the upper limit of typical chip dimensions
- Signal reception covering the entire biomass storage site, provided by multiple receivers with filtering of duplicated measured data
- Network communication—secure communication channels using SSL encryption
- Platform-independent user-friendly client software

4.2.1   The Hardware Part of the System

The hardware part of the monitoring system consists of the required number of measurement probes transmitting the measured data and one or more receiving modules covering the monitored site.

RTM Measurement Probe The measurement probe module (RTM module described in Sect. 2.5.2) is designed to monitor the target environmental parameters such as temperature, humidity, atmospheric pressure, solar radiation intensity, and others. For the temperature monitoring in biomass, a variant containing a temperature sensor has been made (Fig. 2.43). The task of the measuring probe is to process the measured value, prepare the telemetry data, format the data, and send the processed data via a radio interface. The mechanical design of the probe took into account the criteria defined by biomass processors, i.e., autonomous operation for a minimum of 3 months, minimum physical dimensions of the probe, robust design resistant to pressure and moisture of the stored biomass, and modular design allowing the replacement of the sensor module with a type serving other physical variables.

RRM Receiver Module The receiver module (Fig. 2.21) is designed for the wireless reception of data sent by the measuring probes. The description of the module is given in Sect. 2.2. The mechanical design of the RRM module housing is in the form of an aluminum box, which is suitable for outdoor installation. The measuring module also includes an external antenna for the reception of the signal at 433 MHz. Figure 4.8 shows the mechanical design of the RTM transmitter and the RRM receiver. Each RRM module can receive data from up to 127 RTM probes.

Figure 4.9 shows the woodchip storage of a company that supplies heat to homes. The photo is from the sensor system installation period when the storage began to fill its capacity. A total of 3 receiving modules (marked by the red crosses on the pillars in the picture) were installed at the site.

Fig. 4.8 RRM measuring module and RTM sensor module

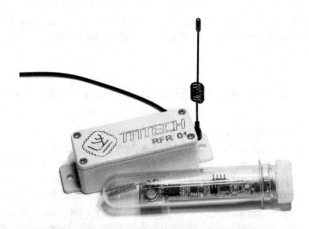

Fig. 4.9 Installing RRM modules in the storage

User Application The measured data is displayed in the nSoric Aurela application. In active monitoring mode (Fig. 3.11), the application works so that the currently measured data is always displayed in the overview window. If any measured value exceeds the allowed value, the application displays a visual notification of this event. In the time record mode (Fig. 3.12), it is possible to display the history of the measured data for the selected location and time interval.

Discussion The test of the measurement system in practice has demonstrated its full functionality and efficiency while complying with all the criteria defined by the biomass processors. The testing was carried out in the woodchip heat producer's plant, with temperature measurement sensors placed in the woodchip at a depth of 2–5 m. Under the laboratory conditions, the probe readout was 0.43 mA in sleep mode and 20 mA in active mode. A long-term power consumption test of the measurement probes showed an average consumption of 0.463 mA, which, when using a standard 2000 mAh AA battery, allows a probe lifetime of at least 180 days. The above parameters, combined with the system's modularity, allow it to be used in various sectors of the national economy, whether in agriculture, food processing, or industry. Monitoring the condition of stored commodities, such as food products, compound feed, and others, appears to be effective.

4.3 Production Hall Temperature Monitoring

Collecting and analyzing the measured data is an essential task in the industry. We can discover seemingly non-existent dependencies between different variables by analyzing measured data. By monitoring the measured variables, we can also predict their future development. By measuring and analyzing the temperature data, we can find out possible causes of heat leakage, identify the parameters of the environment we are measuring and alike. The problem presented in this case study is estimating the temperature at all points of an imaginary network based on temperature data measurements in a particular area where the temperature sensors are distributed mostly irregularly. Once this problem is solved, it is possible to create a dynamic visualization that will contain such estimates in time in sequence [5].

4.3.1 Technical Solution

The sensor system was installed on the premises of a manufacturing company in the Slovak Republic. The installed infrastructure covers three production halls and several dozen office and common areas. The installed system has the following parameters:

- 5 OWB modules were used.
- The total length of the RS485 communication bus was approximately 1000 m.

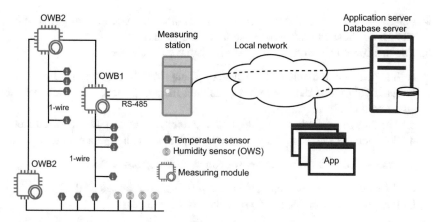

Fig. 4.10 Principal wiring diagram of the system for monitoring the environmental parameters of the production hall

- From 20 to 60, DS18B20 temperature sensors were connected to each OWB module (Fig. 2.11). The total number of temperature sensors was 120.
- 10 OWS modules with relative humidity measurements were used in the installation
- 1-wire bus length for one module was up to 500 m. The total length of the 1-wire bus was approximately 2000 m.

The principal wiring diagram is shown in Fig. 4.10. As this was an installation in a commercial manufacturing company with a rather strict security policy, the system had to be placed in an internal network. The measured data was stored in the same way described in Sect. 3.2.2. Access to the data was also only possible from the internal network.

4.3.2 Software Solution

The specific feature of the installation of the sensor system in the premises of the mentioned company was that the software solution was developed according to the client's exact requirements. The nSoric IS module for storing the measured data and the nSoric API for data access were used as a basis. The user software that was developed was named nSoric Salix. The following functional features characterize it:

1. It uses the nSoric API module to communicate with the IS.
2. Authorization is required to access the data.
3. Existing sensors in the sensor network can be organized into logical sets at two levels: area and sector.

4. For each sensor, the alarm levels can be defined: a lower and an upper value; when a notification of this state should be displayed.
5. The measured data from each sensor is displayed in a table.

   a. It is possible to display all the available data from the measurement history.
   b. Each column of the table represents data from one sensor; the name of this column can be modified.
   c. The data in the table are highlighted in color according to the measured value.
   d. For each measured physical quantity, the data are in a separate table.

6. The measured data from each sensor can be visualized in a graph.

   a. When displaying the graph, the desired sensors can be selected.
   b. It is possible to define the time interval of data display.

7. The measured data can be exported to CSV format.
8. The physical location of the individual files is represented in the situation diagram of the system.

   a. The temperature change over a defined time interval is available in the form of a temperature map animation over the situational diagram of the monitored area.

9. Creation of email notifications when the allowed sensor values are exceeded:

   a. Define and manage the users to whom the notification is sent
   b. Defining the set of sensors for which notifications are sent

Due to the nature of the application and to meet the agreed deadlines and required functionality, the non-functional requirements were defined prior to software development:

1. The application is programmed in Python programming language. In fact, during the development, it was assumed that the nSoric Salix application would be continuously modified. The Python programming language offers many existing modules and is useful for rapid prototyping, which makes it a perfect fit for our purposes.

In Fig. 4.11 is a preview of the nSoric Salix software window. The image shows the color highlighting of the table cells in case the values have dropped below the lower allowed limit. The logical arrangement of the sensors by region can be seen in the upper right-hand part.

An essential part of the software is the display of time records for selected sensors (Fig. 4.12). Before plotting the graph, selecting the sensors to be displayed and the displayed interval is necessary.

A specific feature of the nSoric Salix software is the mapping of the sensor according to its physical location in the situational diagram of the monitored area. Figure 4.13 shows a situational map of the monitored area with a visualization of the measured temperature level.

	Main corr dow	TEST corr	TEST OP	TEST ENT	DUR old	CLIM	BR1	BR2	MAT L1	FLAMM	SR 1	ACC	SR 2	WS off	WS	ATC	ATC off	ATC L	DUR new	PT JIT
00:30	14.1	17.1	21.4	18.2	21.8	23.0	22.6	22.4	21.7	22.8	19.9	20.7	20.6	22.1	21.8	21.6	22.3	22.2	21.4	21.2
00:45	14.0	16.9	21.4	18.1	21.8	23.0	22.6	22.4	21.7	22.8	19.9	20.7	20.6	22.0	21.8	21.6	22.2	22.1	21.3	21.2
01:00	14.0	16.9	21.4	18.1	21.6	22.9	22.6	22.3	21.6	22.6	19.9	20.6	20.5	21.9	21.7	21.5	22.2	22.1	21.2	21.1
01:15	13.9	16.8	21.4	18.1	21.6	22.9	22.6	22.3	21.6	22.6	19.9	20.6	20.5	21.9	21.8	21.5	22.2	22.1	21.4	21.1
01:30	13.9	16.8	21.4	18.0	21.6	22.9	22.6	22.2	21.6	22.5	19.9	20.6	20.5	21.9	22.4	21.5	22.2	22.1	21.6	21.3
01:45	13.9	16.8	21.4	18.0	21.5	22.9	22.5	22.2	21.5	22.6	19.8	20.5	20.4	22.1	22.8	21.5	22.2	22.1	21.8	21.8
02:00	13.9	16.8	21.4	17.9	21.4	22.8	22.5	22.2	21.5	22.4	19.8	20.5	20.4	22.2	22.9	21.5	22.1	22.1	21.9	21.8
02:15	14.0	16.8	21.3	17.9	22.8	23.4	22.5	22.2	21.5	22.4	19.8	20.5	20.4	22.2	22.9	21.5	22.1	22.1	21.9	21.9
02:30	14.0	16.8	21.3	17.9	23.4	23.7	22.4	22.2	21.5	22.4	19.8	20.4	20.4	22.2	23.0	21.5	22.1	22.1	21.9	21.9
02:45	13.9	16.8	21.3	17.9	23.6	23.8	22.5	22.2	21.5	22.4	19.8	20.4	20.4	22.3	23.1	21.5	22.1	22.1	22.0	22.0
03:00	13.9	16.8	21.3	17.9	23.7	23.9	22.5	22.2	21.5	22.4	19.8	20.4	20.3	22.4	23.1	21.5	22.1	22.1	22.0	22.1
03:15	13.9	16.8	21.4	18.0	23.8	23.9	22.5	22.2	21.5	23.1	19.8	20.4	20.3	22.3	23.1	21.5	22.1	22.1	22.0	22.1
03:30	14.0	16.9	21.4	18.0	23.8	23.9	22.5	22.2	21.5	22.7	19.8	20.4	20.3	22.4	23.1	21.5	22.1	22.1	22.0	22.1
03:45	14.1	17.0	21.4	18.1	23.9	24.0	22.5	22.2	21.4	22.6	19.8	20.4	20.3	22.4	23.1	21.5	22.1	22.1	22.1	22.2
04:00	14.2	17.1	21.4	18.1	23.9	24.1	22.5	22.2	21.5	22.5	19.8	20.4	20.3	22.4	23.1	21.6	22.1	22.1	22.1	22.2
04:15	14.2	17.0	21.4	18.1	23.9	24.0	22.5	22.2	21.5	22.4	19.8	20.4	20.2	22.4	23.2	21.6	22.1	22.1	22.1	22.2

Fig. 4.11 nSorix Salix software GUI

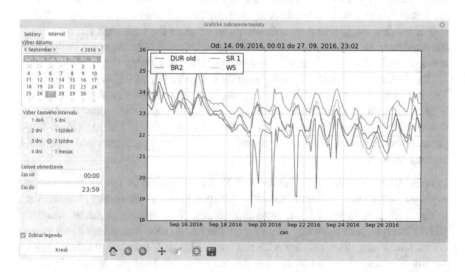

Fig. 4.12 Graphical representation of the measured temperatures

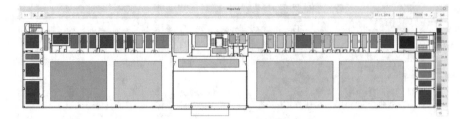

Fig. 4.13 A temperature map for the selected area

In the proposed nSoric Salix software, the visualization of temperature maps is implemented straightforwardly—the value from a specific sensor is displayed using the color scale. In contrast, this measured value's validity area is defined according to its location or defined area. For a more illustrative representation of the overall temperature map, a mathematical model must be created to calculate the temperature values in places where it was not measured. Such an approach assumes that the sensors are placed in an area with no obstacles preventing the heat transfer between parts. Such obstacles are, for example, walls or otherwise separated parts of the monitored area. This approach can provide a more visual representation of the temperature map in some monitored areas. In Fig. 4.13, there are two such areas: in the figure's left- and right-hand parts; there are always two monitored local areas.

For solving the problem of estimating the value of the measured temperature at the monitored area, depending on the known measured values, there are well-known methods such as Inverse Distance Weighting (IDW) or Kriging [6].

The primary parameter of these methods is the density of the resulting network. The individual methods differ in their complexity and especially in their accuracy. When comparing the resulting temperature with the estimated value, the Kriging model appears to be the more suitable since the weights of the individual input points (respectively, of the measured temperatures) can be parameterized depending on their location.

Inverse Distance Method The Inverse Distance method is straightforward and gives us the resulting estimate quickly. However, in this method, any characteristic function cannot influence the resulting values. The inverse distance weighted (IDW) interpolation explicitly implements the assumption that things close to one another are more alike than those farther apart. The IDW uses the measured values surrounding the prediction location to predict a value for any unmeasured location. The measured values closest to the prediction location have more influence on the predicted value than those farther away. The IDW assumes that each measured point has a local influence that diminishes with distance. It gives greater weights to points closest to the input (known) points, and the weights diminish as a distance function. Hence the name inverse distance weighted [5].

The basic relationship for calculating one value per area of interest is

Fig. 4.14 Temperature map
used Inverse Distance method
to compute it [5]

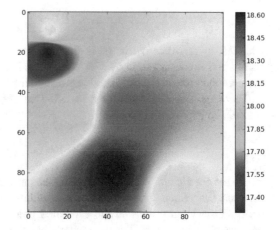

$$Z_0 = \frac{\displaystyle\sum_{n=1}^{n} \frac{z_i}{d_i^p}}{\displaystyle\sum_{n=1}^{n} d_i^p} \tag{4.1}$$

where Z_0 is the estimated value, z_i is the known value, and d_i is the distance between Z_0 and $_zi$. Using the parameter p, one can determine the weight of each reference value. N is the number of known values.

Figure 4.14 shows the temperature map obtained using the Inverse Distance method for 6 input values. The resulting network has a density of 100×100 points.

Kriging Method Kriging is based on the assumption that the interpolated parameter can be treated as a regionalized variable. A regionalized variable is intermediate between a truly random variable and a completely deterministic variable in that it varies continuously from one location to the following. Therefore, points near each other have a certain degree of spatial correlation, but widely separated points are statistically independent. Kriging method represents a set of linear regression routines which minimize estimation variance from a predefined covariance model [5].

The fundamental relationship for estimating the value is

$$Z_0 = \sum_{n=1}^{n} \lambda_i Z_i \tag{4.2}$$

where Z_0 is the estimated value, Z_i is the known value, and λ_i is the weight of the i-th value. The power of this method lies precisely in calculating the individual λ_i weights. The individual weights depend on the distance of the values from which we estimate the value. The following matrix equation applies to calculate the weights:

$$\Gamma \lambda = \gamma \qquad (4.3)$$

where Γ is an $n \times n$ matrix (n is the number of known values) expressing the correlation between the known values and γ is a vector of size n expressing the correlation between the estimated value and all the known values.

There are several models for calculating the correlations (Γ matrix). In our case, we used an exponential model:

$$C(h) = ce^{\frac{-3h}{a}} \qquad (4.4)$$

where the coefficients c and a depend on the particular application, h is the distance between the values. The parameter a specifies a threshold value of the distance between the input point and the estimated point, beyond which the weight value changes only minimally. The parameter c has a similar meaning, but it is applied to all the weights equally.

In our case, based on empirical results, we chose $c = 10$ and $a = 10$. After calculating the values of the Γ matrix and the γ vector, we can calculate the λ weights. However, before that, we need to modify relation 4.3 to ensure that the following condition is met:

$$\sum_{i=1}^{n} \lambda_1 = 1 \qquad (4.5)$$

so that the sum of all the weights is equal to 1. From relation 4.4, we express the lambda weights:

$$\begin{bmatrix} \lambda \\ \mu \end{bmatrix} = \begin{bmatrix} \Gamma & 1 \\ 1 & 0 \end{bmatrix}^{-1} \begin{bmatrix} \gamma \\ 1 \end{bmatrix} \qquad (4.6)$$

To calculate the unknown value, we use relation 4.2. Figure 4.15 shows the result of estimating the values in the study area. In addition, the same temperature input values as in the first case were used in this example.

The covariance function can be used to influence the correlation between the input values and the resulting estimated values. There are several types of standard covariance functions. Relationship 4.4 is an exponential model. Relationship 4.7 defines a spherical model.

$$C(h) = \begin{cases} \left(0.5 \left(\frac{h}{a}\right)^3 - 1.5\frac{h}{a}\right) + 1 & h \leq a \\ 0 & h > a \end{cases} \qquad (4.7)$$

The advantage of using the covariogram is that its input points weight values can be influenced depending on where the value we want to estimate is located. Thus, in

Fig. 4.15 Temperature map
used Kriging method [5]

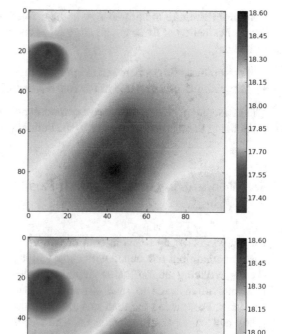

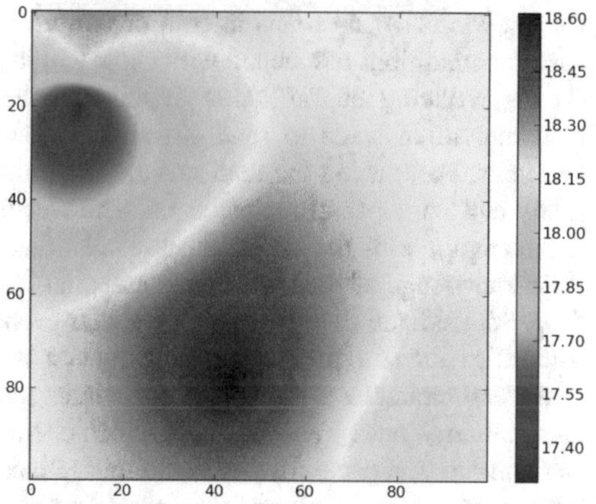

Fig. 4.16 Temperature map
used Kriging
method—modified
covariogram [5]

addition to distance, the direction can also affect the resulting weights (Fig. 4.16). Such a covariogram is then called directional.

Discussion The nSoric monitoring system was used as a supplemental system to control the efficiency of the existing heating, ventilation, and air conditioning (HVAC) system. The client needed to identify locations with sub-optimal temperature conditions in relation to the required temperatures to ensure the necessary comfort for the employees at work and the right conditions for specific technological processes. The first important step was the installation of approximately 120 sensors. Using 1-wire technology reduced the complexity of installing the communication buses to a minimum. When monitoring the condition of the monitored area, the display of the measured data in tabular form, with color highlighting values outside the acceptable range of values, was of great help to the software operator. A graphical representation of the values—time plots of the measured values—can provide similar information. A significant contribution to overall energy savings and HVAC system optimization is the temperature maps, which show the approximated state over the entire monitored area in the defined density of the resulting network. Proper evaluation of these temperature maps contributed to the identification of the following:

- Temperature bridges between the production halls and connecting corridors
- Temperature losses related to the construction of the monitored building

This consequently led to the necessary modifications of the structural elements, thus minimizing the thermal losses due to the building structure itself.

4.4   Monitoring of the Asphalt Road and Its Subgrade Freezing

During winter, many roads in cold or temperate regions are influenced by severe climatic conditions associated with snow and ice. These conditions seriously affect driving, reducing the traffic flow dramatically because the failure to maintain roads in winter often leads to road closures. Ice and snow also increase the risk of accidents. Forecasting the road surface and traffic conditions is essential for traffic safety and winter road maintenance. The weather conditions can change quickly, for example, with the onset of snowfall or during rapid temperature variations. Prior knowledge of road weather is essential from a public road safety standpoint. Proper consideration of upcoming weather events also helps the road maintenance authorities attend to the roads effectively and economically [7].

Several strategies can be used for winter road maintenance. Among these, the most common ones are de-icing, in which chemicals are used to melt ice and snow, and anti-icing, a preventive measure that reduces ice by hindering bonds between ice crystals and road pavement or spreading sand to help provide traction [8]. The most used de-icing chemical is salt (sodium chloride, NaCl), which usually comes from mined rock salt. It is one of the most cost-effective, readily available, and efficient chemicals for the prevention and removal of ice and snow, is supplied in various granule sizes, and is occasionally applied in solution. Traditionally, this has been spread in dry form, known as dry salt. The impact of using chemicals for winter road maintenance is a primary environmental concern. Studies [9, 10] show that soils, vegetation, water, highway structures, and vehicles are all affected, so the de-icing chemicals must be used wisely. The growing awareness of the environmental impacts of salt has led to efforts to find ways to reduce the amounts entering the environment. There are only two ways to achieve this. They are using something other than salt or optimizing salt usage by applying it strategically. Hence, there is a continual need for improvement in winter road maintenance strategy. One key component of optimized road maintenance decisions is to obtain and use accurate weather information. Weather information can be divided into two temporal categories: observations, which reflect current conditions, and forecasts, which predict future conditions. The most considerable potential savings to be made in winter maintenance focus on predicting ice formation and snow. Countries prone to icy roads can make significant annual savings [11] by optimizing the weather forecasts with salt usage and other operational costs.

Thus, it stands to reason that it is necessary to have relevant data about the weather and road surface conditions. They can be obtained from various sources. The most essential real-time meteorological and pavement data source is the road weather stations (RWSs), ensuring automatic data collection and transmission. Currently, a comprehensive range of technologies for road weather stations are commercially available (e.g., Vaisala, Lufft, CrossMet, and others). They are an essential tool for remote monitoring of road weather conditions. The RWS is equipped with various sensors for measuring meteorological data. They are also equipped with road sensors connected to the station to measure road conditions. Road surface conditions are influenced by numerous meteorological, geographical, and road parameters, which can produce vast temperature variations across the road section. The most critical factors are air temperature, radiation fluxes, humidity, precipitation amount, wind, topography, properties of road materials, and traffic [12, 13]. The literature presents their influence on the road in detail [14].

These RWSs are a part of advanced systems known as road weather information systems (RWISs) and maintenance decision support systems (MDSSs). The RWIS is a software tool that collects all the relevant available data that dispatchers can use to support decision-making in winter road maintenance management. The RWIS integrates current data from road weather stations, meteorological radars and satellites, cameras, mobile road weather stations, the National Weather Service (NWS), and others. The RWS ensures the recording of current conditions in selected locations, but the RWIS can include a forecasting module, allowing each RWS to generate a point forecast of essential parameters in response to the measured data and general weather forecast. The NWS generally supplies general weather forecasts, alternatively by a commercial provider. In addition, there are specialized computational models for specialized forecasts for road surfaces. The specialized road weather forecast model is a part of the dispatcher's maintenance decision support system (MDSS). The MDSS, as a part of the RWIS, is focused on particular linear weather and road surface status forecasts and recommendations on winter maintenance [15]. More information about RWIS and MDSS is presented in [15]. Various models for predicting meteorological conditions on the road have been developed (e.g., METRo [16], RoadSurf [7] and others). Based on these models' forecast road condition information, the highway engineer can determine the window of application for anti- and de-icing chemicals and for snow removal operations to commence. By strategically timing operations, the safety of highways can be maintained while going some way to optimizing the quantity of material applied to the road surface.

Here we present the design and implementation of an experimental system in which the desired value of the environment, such as the temperature of the road, is obtained by the direct method, i.e., sensors are placed directly into the road surface. Given this data acquisition method, the model is suitable for use in local areas. The data are more accurate than measuring the air temperature using the weather station near the road.

4.4.1 Experimental Installation

An experimental installation of the temperature sensors, buried in the road surface, was created to verify the model described in the next section. The sensors are divided into two groups: the horizontal layer, located in the road top layer, and the vertical layer, where the sensors are arranged in the vertical direction with a pitch of 20 cm to a depth of 1 m. Figure 4.17 illustrates the location of temperature sensors in the experimental installation. Temperature sensors were made in a special enclosure that protects them against mechanical damage during installation and subsequent use.

The following restrictions of the measurement system were defined for the experimental installation:

- A robust mechanical enclosure of the temperature sensors to withstand the high temperature of asphalt installation and mechanical stress during the operation of the road in which the sensors are located.
- Minimized cabling. Since the system contains more than 10 temperature sensors, it is required to minimize the cabling.
- The temperature measurement ranges from—50 °C to 120 °C.
- Simple interface through which it is possible to view and evaluate the measurement results.

The requirement for the robust mechanical design was met in the sensor enclosure (Fig. 4.19) made of Ertalon PA66, having the following essential characteristics: melting point 255 °C, thermal conductivity 0.28 W/(K.m), the short-term operating temperature of 200 °C, and the long-term operating temperature of 80 °C.

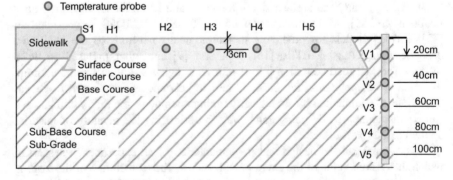

Fig. 4.17 Design of sensors placement in the asphalt road [17]

Fig. 4.18 Horizontal temperature sensor design

Fig. 4.19 Installation of temperature sensors in asphalt roads

The OWB measurement module was used for the installation, to which 11 sensors were connected in a unique mechanical design. The sensors embedded in the body of the asphalt pavement (Fig. 4.17, sensors H1 to H5) are encased in a special housing. The sensor itself has contact with the environment at its top part. Figure 4.18 shows an authentic detail of this sensor. The protective housing for the wiring can also be seen in the figure. The installation of these sensors in asphalt road is shown on Fig. 4.19.

Figure 4.20 shows a record of temperature measurements over 35 days. The graph shows the values from the vertically stacked sensors. From this plot, the phase shifts between the curves, i.e., the heat transfer in the environment, can be inferred.

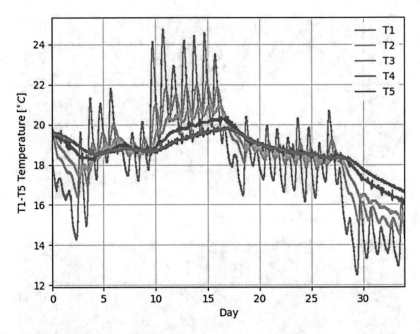

Fig. 4.20 Measured temperatures from vertical sensors

With long-term observation, the soil profile's rate of freezing or heating (thawing) at different depths can be seen.

4.5 Experimental Measurement of Asphalt Pavement Fatigue

The primary sources of internal deformation in the pavement structure are induced by the action of heavy vehicles on the pavement. Due to the complexity of the interaction, a combination of experimental measurements in the laboratory and computational simulation is considered the preferred approach. The sensing device embedded in the pavement, tested with the linear semi-mobile APT Tester 105-03-01 [18], provided sufficient data during 5 years of operation. This data was used to develop a numerical model that describes the interaction between the rear axle of the heavy vehicle and the roadway.

The APT Tester 105-03-01 (Fig. 4.21) emulates the passage of a heavy vehicle axle on an asphalt road. The device itself consists of an asphalt roadway (Fig. 4.22) with a size of 3m x 6m loaded by one heavy vehicle axle. The axle contact force and the speed at which the axle moves are configurable and set to match the actual situation.

Figure 4.22 shows the axis of motion of a half-axle with two tires. Heavy trucks have the same design. The dashed line represents the axis of motion for

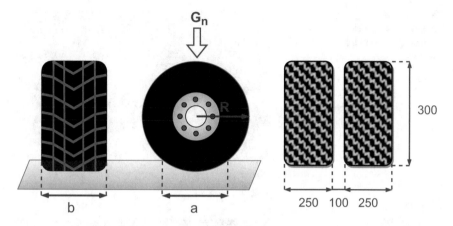

Fig. 4.21 Axle sketch of semi-automatic APT Tester 105-03-01 [19]

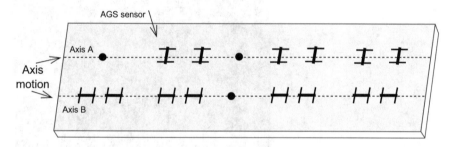

Fig. 4.22 Placement of the AGS sensors in the body of the asphalt road [19]

each half-axle tire. The AGS sensors were placed in each axis of motion in a different orientation. The sensor orientation is essential since the AGS sensor only measures deformation in the longitudinal direction, as indicated in Fig. 4.23. Sensors located in the A-axis detect the deformation of the asphalt in the transverse direction concerning the direction of the wheel motion. The sensors in the B-axis detect the deformation in the direction of wheel motion. The measured deformation in the B-axis is assumed to be more prominent during acceleration and braking.

In Fig. 4.24 is shown the physical embedding of the AGS sensors in the asphalt road. The figure is from the penultimate stage of the installation when the AGS sensors were inserted into the prepared position. The last phase of the installation was to apply the final layer of asphalt where the sensors were placed and to level the surface using an industrial vibratory asphalt roller.

Fig. 4.23 AGS sensor with the representation of measured deformation direction

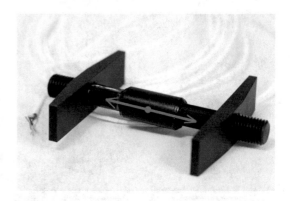

Fig. 4.24 Installation of AGS sensors in asphalt road

4.5.1 Deformation Measurement

TNZ measurement modules were used for the measurements of the AGS sensors as a part of the verification and calibration phase of the installation. A total of 14 AGS sensors were installed. During this verification phase, verification measurements were taken regularly to establish a steady state. For the measurements themselves, the output was a value that corresponded to the strain gauge deflection. The strain gauge is a part of the AGS sensor body, and the sensor's deformation depends on the force applied in the strain axis of the sensor (Fig. 4.23). This strain force causes a change in the length of the sensing part of the sensor. The measurements in the calibration phase resulted in the determination of an engineering zero—i.e., the value of the measured strain at which the zero external force is applied.

The relation 4.8 defines the unit for expressing the strain. Deformation is defined as the response of a system to applied stress. Stress is generated when a substance or material is loaded by force. Stress causes the deformation of that material. Deformation is defined as the amount of material deformation in the applied force's direction divided by the material's initial length.

$$strain(l) = \frac{\Delta l}{l} \tag{4.8}$$

where Δl is the change in length of the measured element and l is its length when no external force is applied.

Figure 4.25 shows the waveform of the measured values recalculated based on the stiffness modulus for the material used and the ambient temperature at the

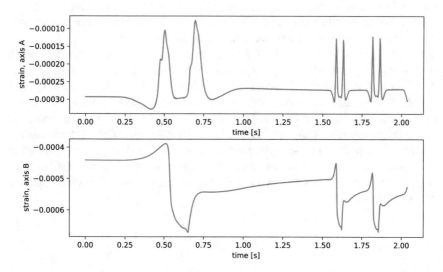

Fig. 4.25 The waveform of the measured relative deformation when dynamic loads are applied

moment of measurement. The plots in Fig. 4.25 show one slow load unit transition followed by two faster transitions. An evaluation of this experiment is presented in [19].

The proposed AGS sensors formed an indispensable role in this experiment. The measured data made it possible to determine the actual state of mechanical stresses in the road body. Those data were subsequently used for correction of the numerical model used by the researchers of the Faculty of Civil Engineering, University of Žilina, Slovakia. The TNZ measurement module was used to calibrate the sensors, and then a measurement station with a sampling frequency of 50 kHz was used to measure the pavement deformations. Despite the low sampling rate of the TNZ measurement module, this module is helpful in processes with a slow change of the monitored variable. Examples of such applications are, e.g., the measurement of mechanical stress in prestressed structures or otherwise permanently loaded systems where the overall trend of the change in mechanical stress is of interest. The TNZ measurement module contains two independent channels that can be configured for strain gauge measurements or precise temperature measurements with the PT100/PT1000 sensor. In both cases, it is a bridge connection.

4.6 Authors' Protected Intellectual Property

This chapter presents the author's design patterns and utility models related to the presented measurement and sensing modules. The sensor for measuring mechanical deformation of asphalt pavement is presented first. The sensor is specific in its mechanical design and ability to withstand high temperatures when installed in hot asphalt. The second utility model relates to a universal serial bus device for measuring physical quantities. This device is a universal hardware module communicating over a 1-wire bus, which can contain any sensor. A third utility design is a sensor for measuring the temperature profile of the asphalt pavement. It is a specially designed temperature sensor for installation in an asphalt road. The fourth design pattern is system for wireless measurement of environmental variables in biologically active materials. This design is a monitoring system based on RF communication, and its use of measurement probes is primarily intended for use in bulk and biologically active materials. The system consists of the necessary measurement probes and several receiving modules. The last protected sensor is the Rod probe for temperature measurement with an adjustable position of installed sensors. This solution is a way of housing temperature sensors with defined spacing between the sensors.

4.6.1   Sensor for Measuring Mechanical Deformation of Asphalt Pavement

Application number   SK932014U1
Utility model number   SK7128Y1

Inventors:

- Fabo Peter
- Ďuďák Juraj
- Dvorský Ludovít

Description:
A sensor for measuring the mechanical deformation of an asphalt pavement comprises a measuring body (1), at least two fixing bodies (2), at least two measuring elements (3), a protective sleeve (4), wiring (6) and a cable protector (5), a thermometer (7), and an electronic identification unit (8). The measuring body (1) is made of a plastic material with material properties close to the installation environment, capable of withstanding asphalt laying temperatures. It is mechanically machined into a shape to allow the connection of the wiring (6) and the cable gland (5) so that the measuring elements (3), the thermometer (7), and the electronic identification unit (8) can be fixed to the surface of the measuring body (1). The fixing bodies (2) are made of the same material as the measuring body (1). The measuring elements (3), the thermometer (7), and the electronic identification unit (8) are fixed to the surface of the measuring body (1) and are connected to the cabling (6). Mechanical protection of the measuring elements (3), the thermometer (7), and the electronic identification unit (8) is provided by a protective sleeve (4) fixed to the measuring body (1) employing a flexible high-temperature sealant (Fig. 4.26). A cable protector (5) provides mechanical protection for the cabling (6) during installation [20].

Fig. 4.26 Sensor for measuring mechanical deformation of asphalt pavement [20]

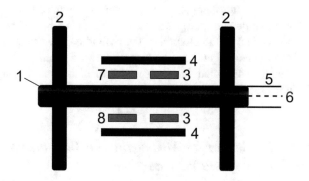

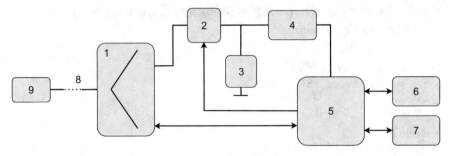

Fig. 4.27 Universal serial bus device for measuring physical quantities [21]

4.6.2   Universal Serial Bus Device for Measuring Physical Quantities

Application number   SK862017U1
Utility model number   SK8225Y1

Inventors:

- Fabo Peter
- Ďuďák Juraj
- Sládek Ivan

Description:
A universal serial bus device for measuring physical quantities allowing the use of any sensors without the need for an external power supply is connected to the serial bus (8) and contains a splitter (1) of the communication and power supply part used for detection of the working and resting state. In the quiescent state, charging of the temporary power source (3) is activated. In the event of detection of a transition to the working state, charging of the temporary power source (3) is suspended by the charging controller (2). Communication with the connected sensors (6) is controlled by a low-power microprocessor (5), which is powered by a supply voltage regulator (4). Mutual differentiation of the individual devices connected to the serial bus (8) is implemented utilizing a unique hardware identifier (7) - Fig. 4.27. Data obtained from the connected sensors (6) can be sent to the serial bus (8) for further processing in the master computer (9) [21].

4.6.3   Sensor for Measuring the Temperature Profile of the Asphalt Pavement

Application number   SK492015U1
Utility model number   SK7348Y1

Fig. 4.28 Sensor for
measuring the temperature
profile of the asphalt
pavement [22]

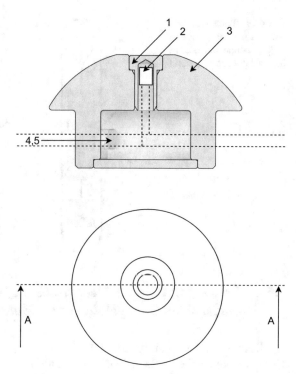

Inventors:

• Gašpar Gabriel

Description:

The sensor for measuring the temperature profile of an asphalt roadway includes a housing (3), a measuring body (1), a digital thermometer (6) with an electronic identification unit (2), wiring (5), and a cable protector (4). The housing (3) is made of a material with properties close to the installation environment and resistant to asphalt laying temperatures (Fig. 4.28). It is mechanically machined into a shape to allow the connection of the wiring (5) and the cable gland (4) so that the measuring body (1) and the digital thermometer with the electronic identification unit (2) can be mounted in it. The measuring body (1) with the embedded thermometer (6) with the electronic identification unit (2) is fixed in the housing (3) and is connected to the cabling (5). A cable protector (4) provides mechanical protection for the cabling (5) during installation [22].

Fig. 4.29 System for wireless measurement of environmental variables in biologically active materials [23]

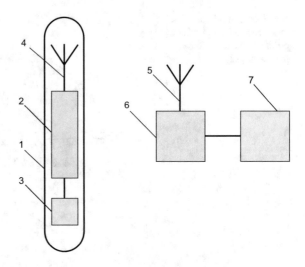

4.6.4   System for Wireless Measurement of Environmental Variables in Biologically Active Materials

Application number   SK1972015U1
Utility model number   SK7657Y1, CZ30330U1

Inventors:

• Gašpar Gabriel

Description:
System for wireless measurement of environmental variables in biologically active materials includes a transmitter (1) comprising a housing resistant to mechanical, chemical, and environmental influences with a label location inside the housing, electronics (2) for measuring and transmitting measured data, environmental sensors (3), and a transmitting antenna (4). The system further comprises receivers comprising a receiving antenna (5), electronics (6) for receiving and processing the data, and a unit (7) for processing and presenting the measured data [23] (Fig. 4.29).

4.6.5   Rod Probe for Temperature Measurement with an Adjustable Position of Installed Sensors

Application number   SK122021U1
Utility model number   SK9323Y1

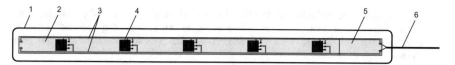

Fig. 4.30 Rod probe for temperature measurement with an adjustable position of installed sensors [24]

Inventors:

- Šedivy Štefan
- Gašpar Gabriel
- Fabo Peter
- Farbák Matúš

Description:

Rod probe with an adjustable position of installed sensors for temperature measurements (Fig. 4.30). The probe consists of plastic housing (1) with dimensions according to the requirements of the specific application in the installation environment, in which one or more modular pads (2) and one termination pad (5) are inserted, which are conductively connected and on which temperature sensors (4) are arranged as required. From the termination pad (5), wiring (6) is routed through the hermetically sealed housing (1) for connection to the measuring device [24].

References

1. J. Ďuďák, I. Sládek, M. Skovajsa, Utilization of nSoric system for monitoring ecological building insulations, in *MMK 2015 - MEZINÁRODNÍ MASARYKOVA KONFERENCE PRO DOKTORANDY A MLADÉ VĚDECKÉ PRACOVNÍKY*, vol. VI (2015). ISBN: 978-80-87952-12-2
2. S. Badurová, I. Sládek, J. Ďuďák, M. Skovajsa, P. Fabo, Ekologické tepelné izolácie a ich aplikácia pri obnovách stavieb, in *Zborník príspevkov z konferencie Research Forum 2015, Vysoké Tatry* (2015). ISBN: 978-80-554-0973-3
3. S. Badurova, J. Jost, F. Bahleda, J. Ďuďák, Analysis of the internal insulation of renovated building. Appl. Mech. Mater. **824**, 363–370 (2016). https://doi.org/10.4028/www.scientific.net/AMM.824.363
4. M. Skovajsa, P. Fabo, L. Pepucha, I. Sládek, Proposal of a wireless measurement system for temperature monitoring of biological active materials, in *Advanced Mechatronics Solutions*, ed. by R. Jablonski, T. Brezina, (Springer International Publishing, 2016), pp. 367–372. ISBN: 978-3-319-23923-1
5. J. Ďuďák, S. Pavlíková, G. Gašpar, Methods of temperature estimation on given area in system of data collection, in *14th International Conference Mechatronika*, pp. 39–42 (2011). https://doi.org/10.1109/MECHATRON.2011.5961088
6. S. Samanta, D. Kumar Pal, D. Lohar, B. Pal, Interpolation of climate variables and temperature modeling. Theor. Appl. Climatol. **107**, 35–45 (2012). https://doi.org/10.1007/s00704-011-0455-3. Accessed: 2022-10-15

7. M. Kangas, M. Heikinheimo, M. Hippi, Roadsurf: a modelling system for predicting road weather and road surface conditions. Meteorological Applications **22**(3), 544–553 (2015). https://doi.org/10.1002/met.1486. https://rmets.onlinelibrary.wiley.com/doi/abs/10.1002/met. 1486. Accessed: 2022-11-19

8. V.J. Berrocal, A.E. Raftery, T. Gneiting, R.C. Steed, Probabilistic weather forecasting for winter road maintenance. J. Am. Stat. Assoc. **105**, 522–537 (2010)

9. D.M. Ramakrishna, T. Viraraghava, Environmental impact of chemical deicers – a review. Water Air Soil Pollut. **166**, 49–63 (2005)

10. H.L. Dai, K.L. Zhang, X.L. Xu, H.Y. Yu, Evaluation on the effects of deicing chemicals on soil and water environment. Procedia Environ. Sci. **13**, 2122–2130 (2012). ISSN: 1878-0296. https://doi.org/10.1016/j.proenv.2012.01.201. https://www.sciencedirect.com/science/article/pii/S1878029612002022. 18th Biennial ISEM Conference on Ecological Modelling for Global Change and Coupled Human and Natural System, Accessed: 2022-11-21

11. A. Kann, G. Pistotnik, B. Bica, INCA-CE: a central European initiative in nowcasting severe weather and its applications. Adv. Sci. Res. **8**(1), 67–75 (2012). https://doi.org/10.5194/asr-8-67-2012. https://asr.copernicus.org/articles/8/67/2012/. Accessed: 2022-10-14

12. R. Krsmanc, A.S. Slak, J. Demsar, Statistical approach for forecasting road surface temperature. Meteorological Applications **20**(4), 439–446 (2013). https://doi.org/10.1002/met.1305. https://rmets.onlinelibrary.wiley.com/doi/abs/10.1002/met.1305. Accessed: 2022-11-05

13. L. Chapman, J. Thornes, Small-scale road surface temperature and condition variations across a road profile, in *Sirwec 2008*, pp. 1–8 (Jan 2008)

14. S. Kawashima, T. Ishida, M. Minomura, T. Miwa, Relations between surface temperature and air temperature on a local scale during winter nights. J. Appl. Meteor. **39**(9), 1570–1579 (2000). https://doi.org/10.1175/1520-0450(2000)039<1570:RBSTAA>2.0.CO; 2. https://journals.ametsoc.org/view/journals/apme/39/9/1520-0450_2000_039_1570_rbstaa_2.0.co_2.xml. Accessed: 2022-11-29

15. A. Kociánová, The intelligent winter road maintenance management in Slovak conditions. Procedia Engineering **111**, 410–419 (2015). ISSN: 1877-7058. https://doi.org/10.1016/j.proeng.2015.07.109. https://www.sciencedirect.com/science/article/pii/S1877705815013582. XXIV R-S-P seminar, Theoretical Foundation of Civil Engineering (24RSP) (TFoCE 2015), Accessed: 2022-12-01

16. S. Linden, S. Drobot, The evolution of metro in a roadway DSS, in *15th International Road Weather Conference* (Feb 2010)

17. J. Ďuďák, G. Gabriel, S. Sedivy, P. Lubomir, Z. Florkova, Road structural elements temperature trends diagnostics using sensory system of own design, vol. 236, 2017. https://doi.org/10.1088/1757-899X/236/1/012036

18. L. Remek, J. Mikolaj, M. Skarupa, Accelerated pavement testing in Slovakia: Apt tester 105-03-01. Procedia Engineering **192**, 765–770 (2017). ISSN: 1877-7058. https://doi.org/10.1016/j.proeng.2017.06.132. https://www.sciencedirect.com/science/article/pii/S1877705817326784. 12th international scientific conference of young scientists on sustainable, modern and safe transport, Accessed: 2022-12-10

19. L. Remek, V. Valaskova, Data gathering and evaluation of tensile strains measured in apt with mathematical computation method, in *Accelerated Pavement Testing to Transport Infrastructure Innovation*, ed. by A. Chabot, P. Hornych, J. Harvey, L.G. Loria-Salazar (Springer International Publishing, Cham, 2020), pp. 448–457. ISBN: 978-3-030-55236-7

20. P. Fabo, J. Ďuďák, L. Dvorský, Senzor na meranie mechanických deformácií asfaltovej vozovky. https://wbr.indprop.gov.sk/WebRegistre/UzitkovyVzor/Detail/93-2014, May 2015. Utility model SK7128Y1, Accessed: 2022-12-12

21. P. Fabo, J. Ďuďák, I. Sládek, Univerzálne zariadenie sériovej zbernice na meranie fyzikálnych veličín. https://wbr.indprop.gov.sk/WebRegistre/UzitkovyVzor/Detail/86-2017, October 2018. Utility model SK8225Y1, Accessed: 2022-12-12

22. G. Gašpar, Senzor na meranie teplotného profilu asfaltovej vozovky. https://wbr.indprop.gov.sk/WebRegistre/UzitkovyVzor/Detail/49-2015, February 2016. Utility model SK7348Y1, Accessed: 2022-12-12

23. G. Gašpar, Systém bezdrôtového merania environmentálnych veličín v biologicky aktívnych materiáloch. https://wbr.indprop.gov.sk/WebRegistre/UzitkovyVzor/Detail/197-2015, January 2017. Utility model SK7657Y1, Accessed: 2022-12-12

24. Š. Šedivý, G. Gašpar, P. Fabo, M. Farbák, Tyčová sonda na meranie teploty s nastaviteľnou pozíciou inštalovaných senzorov. https://wbr.indprop.gov.sk/WebRegistre/UzitkovyVzor/Detail/197-2015, September 2021. Utility model SK9323Y1, Accessed: 2022-12-12

Glossary

1-Wire is a registered trademark of Maxim Integrated Products, Inc., all rights reserved. Maxim Integrated Products Inc. is a wholly owned subsidiary of Analog Devices Inc.

AGS sensor Sensor AGS consists of a longitudinal sensor body on which a strain gauge is installed and two arms that ensure mechanical deformation of the sensor body when force is applied to these arms. The body of the sensor itself can be subjected to tension or pressure.

DBar probe A universally designed printed circuit board with defined spacing for easy installation of DS18B20 sensors. The required length for the vertical sensor can be created by connecting several modules together.

MultiSlave MultiSlave module represents several simple slave devices, which are physically designed as a single device. Since there are multiple slave devices, each slave device must have its address. From the MultiSlave node functionality point of view, the individual modules are connected, for example, via an internal bus.

Measuring module OWB The OWB module represents a 1-wire bus controller with improved features. It contains an improved dynamic pull-up controller, a 1-wire bus length detector, and an automatic setting of 1-wire bus communication parameters, depending on the bus length.

Measuring module TNZ TNZ module (strain gauge measurement module) allowed measurement of bridge connections with application to strain gauge measurements.

Measuring module THM (Temperature Humidity Module) is a module for measurement on four DHTxx-type sensors.

Measuring module RRM (Radio Receiver Module) is receiving module for short-range radio communication. It works with the RTM sensor module.

© The Author(s), under exclusive license to Springer Nature Switzerland AG 2023
J. Ďuďák, G. Gašpar, *Design and Implementation of Sensory Solutions for Industrial Environment*, Signals and Communication Technology, https://doi.org/10.1007/978-3-031-30152-0

nSoric API (Application Programming Interface) It is used for data access in nSoric IS. This interface enables remote diagnostics and configuration of the measurement modules.

nSoric Aurela Desktop software that provides visualization of sensory data by accessing the data stored on a remote server. It uses the nSoric API for communication.

nSoric Cofig Is a configuration software for the measurement modules. It allows the upload of information to the module, such as the module's name, date of installation, used version of firmware and hardware, list of sensors that the module contains, and parameters of these sensors, or the calibration coefficients.

nSoric IS (nSoric Information System) Is a data model of the sensory system. The data model is built on a MySQL relational database.

nSoric Merula Desktop software that provides on-demand measurements on the measurement modules. The data is stored locally on the computer where the software is running.

nSoric senlib A library that provides basic functionality for all the other components. It contains functionality for communication with the nSoric API and locally connected measurement modules.

Sensory module OWS (One-Wire Slave) Is a solution for using any sensor on a 1-wire bus. It primarily cooperates with the OWB measuring module.

Sensory module RTM (Radio Transmitter Module) Is a sensory module with support for sending the measured data via the radio interface. It primarily cooperates with the RRM measurement module.

SuperSlave This module enabled slave/master communication in specific cases. The need for such communication may appear when transmitting certain information (for example, real-time information) to all the slave nodes.

Index

Symbols
1-wire, 185, 191
1-wire bus, 51, 79–81, 83, 85–87, 89, 91, 108,
    117–121, 126–128, 159, 160, 166,
    167
1-wire commands, 58
1-wire communication, 59, 72
1-wire master, 54, 55, 57–61, 63, 67, 68, 70,
    74, 121, 126
1-wire network, 71
1-wire operation, 57, 67
1-wire protocol, 52, 53
1-wire sensor, 48, 50, 51, 54
1-wire slave, 63, 64
3D model, 92, 95

A
Accumulated value, 50
Address field, 5, 12, 177
Addressing algorithms, 68
Addressing rules, 10
Advanced Encryption Standard (AES), 25,
    134
Application Data Unit (ADU), 5, 10, 20,
    26
Application firmware, 38
Application programming interface (API), 78,
    150, 152, 154, 157, 164, 172, 174,
    185
ARM Cortex M0, 23, 82, 170
ARM Cortex M3, 24
Asphalt road, 92, 115

B
Battery, 91, 118, 128, 129, 132, 134, 161, 169,
    184
Bidirectional, 52, 53
Broadcast, 5, 9, 11, 58, 59, 61, 63, 64, 68, 103,
    131

C
Cabling, 2, 3, 72, 194
Checksum, 17, 20, 34, 35, 37, 58, 63
Cipher, 134–137
Cipher key size, 20
Communication bus bridge, 49
Communication encryption, 38, 51
Communication protocol, 77, 79, 99, 101, 103,
    124, 129, 172, 182
Communication reliability, 55
Configuration memory, 28
Connectors, 2, 3
CRC16, 20, 34
Cyclic Redundancy Check (CRC), 6, 34, 37,
    120, 131, 133, 137

D
Data Encryption Standard (DES), 25
Datetime, 154
Dbar probe, 94, 95
Decrypting, 27
Decryption, 135, 137, 139
Destination address, 35, 37
Digital thermometer, 51, 167
Digital thermometer DS18B20, 51, 58

© The Author(s), under exclusive license to Springer Nature Switzerland AG 2023
J. Ďuďák, G. Gašpar, *Design and Implementation of Sensory Solutions for
Industrial Environment*, Signals and Communication Technology,
https://doi.org/10.1007/978-3-031-30152-0

E
EIA-485, 1
Encrypted communication, 23, 29
Error code (EC), 34, 85
ERTALON, the material, 115–117
Ethernet, 164, 168, 169, 171, 173, 174
Exception code (EC), 6
Externally powered, 53

F
Fail-safe bias, 3
Family code, 53, 58, 63
Fieldbus, 17
Function code (FC), 6, 13, 17, 20, 34, 35, 61

G
Grounding, 3

H
Half-duplex, 52, 80, 82
HASH, 25
Humidity, relative air, 103, 104, 106–108, 120,
       128, 132, 159, 185, 193

I
Industrial bus, 17
Interference, 55
Internet of Things (IoT), 168, 169, 171
ISO/OSI, 4

J
JSON format, 148, 152

K
Key length, 24, 136
Keys exchange, 22

L
Linear topology, 55
Little Endian, 35
Long-distance, 52

M
Master, 4–8, 79, 83, 128
Maximum weight, 71
Measurement frequency, 44

Measurement methods, 30
Measurement system, 148, 150–153, 161, 180,
       182, 184, 194, 200
Measuring module, 77, 78, 80, 87, 99, 104,
       108, 115, 118, 133
Measuring probe, 182, 183
Mechanical damage, 194
Mechanical deformation, 115, 117
Mechanical design, 91, 94, 115, 129, 182, 183,
       194, 195
Mechanical stress, 182, 200
Memory map, 28
Microcontroller, 23, 25, 72, 82, 86, 100, 101,
       104, 108, 118, 121, 123, 124,
       126–128, 134, 136, 137, 139
MODBUS, 4, 5, 10, 13, 32
Modulation, ASK, 131
Module address, 28, 34, 158, 160
Module configuration, 48
Module status, 38
Multi-platform, 142
MultiSlave, 10, 11, 78, 100, 104, 109
Multislave module, 27, 29, 36
MySQL database, 141, 143, 145

N
Non-secured communication, 18

O
OWB module, 43, 46, 49, 80, 82, 84, 85, 88,
       89, 160, 166, 167
OWS module, 117–119, 121, 123

P
Padding scheme, 25
Parasitically powered, 53
Physical principle, 30
Physical unit, 30
Power consumption, 54, 121, 123, 124, 126,
       127, 169–171, 184
Power-saving, 121
Presence pulse, 55, 72
Private key, 21, 23, 25
Protocol Data Unit (PDU), 5, 10, 20, 25
Public key, 21, 51
Pull-up, 56, 67, 71, 72, 74, 75, 80, 84,
       89

Q
Q numeric format, 29

R
Radio-Frequency (RF), 99–101, 103, 128, 137
Radio Receiving Module (RRM module), 100,
 101, 104, 133, 159
Radius of the network, 54
Read measured values, 49
Reset, 41, 45, 147
Reset pulse, 58, 63, 64
Resistive terminators, 2
Retransmission, 9
Rewritable memory, 63
RS-485, 77–82, 100, 104, 108, 147, 158, 173,
 174, 177, 180, 184
RSA algorithm, 21, 23, 25, 26

S
Search algorithm, 59, 60, 63, 70
Secure communication, 18
Secure layer, 27, 133, 146, 174
Senlib library, 141, 142, 145, 157, 158
Sensor network, 17, 54, 158
Sensor type, 30, 39
Sensory system, 77, 78, 141–143, 162, 163
Slave, 7, 9
Slave identifier, 18, 63
Slave name, 18
Star topology, 54, 55
State diagram, 7, 12, 14, 58, 61
STM32, 23, 79, 101, 104, 118, 123, 126, 127,
 129, 136, 139
Strain gauge, 80, 108, 111, 114–116
Stubbed topology, 55
Submodule, 12, 28, 36, 41, 46, 49, 50
Sub-slave, 28–30, 36, 47, 48, 50, 100, 104,
 109, 166, 167
Supercapacitor, 83, 119, 120, 123, 124, 126,
 127

Super request (SR), 13, 16
SuperSlave, 10, 13

T
Temperature conversion, 49, 58, 119
Temperature sensor, 92, 93, 111, 117, 120,
 127, 129, 137, 161, 178, 182, 184,
 185, 194
Termination, 2, 72
TNZ module, 109–111, 159, 199, 200
Transmission speed, 80, 82, 131
TripleDES, 25

U
UBUS application protocol, 10, 30, 77, 85, 88,
 114, 146, 147, 158
UBUS error codes, 34
UBUS response format, 36
UBUS unicast functions, 33
Unencrypted, 20, 23
Unicast, 9, 11, 18, 58
Unique ID, 63
Unique identifier, 29, 38
Unshielded twisted-pair, 72

V
Virtual slave, 11

W
Weight of the network, 54, 55
Wireless, 79, 99, 103, 133, 159, 169–172, 183
Wiring diagram, 85, 89, 109, 111–113, 121,
 123

Printed in the United States
by Baker & Taylor Publisher Services